ROUTLEDGE LIBRARY EDITIONS:
URBAN PLANNING

Volume 8

URBAN MARKETS

URBAN MARKETS
Developing Informal Retailing

DAVID DEWAR AND VANESSA WATSON

Routledge
Taylor & Francis Group

LONDON AND NEW YORK

First published in 1990 by Routledge

This edition first published in 2018
by Routledge
2 Park Square, Milton Park, Abingdon, Oxon OX14 4RN

and by Routledge
711 Third Avenue, New York, NY 10017

Routledge is an imprint of the Taylor & Francis Group, an informa business

British Library Cataloguing in Publication Data
A catalogue record for this book is available from the British Library

ISBN: 978-1-138-49611-8 (Set)
ISBN: 978-1-351-02214-9 (Set) (ebk)
ISBN: 978-1-138-48527-3 (Volume 8) (hbk)
ISBN: 978-1-138-48531-0 (Volume 8) (pbk)
ISBN: 978-1-351-04987-0 (Volume 8) (ebk)

Publisher's Note
The publisher has gone to great lengths to ensure the quality of this reprint but
points out that some imperfections in the original copies may be apparent.

Disclaimer
The publisher has made every effort to trace copyright holders and would welcome
correspondence from those they have been unable to trace.

Urban Markets
Developing Informal Retailing

David Dewar and Vanessa Watson

Routledge

London and Now York

First published 1990
by Routledge
11 New Fetter Lane, London EC4P 4EE

Simultaneously published in the USA and Canada
by Routledge
a division of Routledge, Chapman and Hall, Inc.
29 West 35th Street, New York, NY10001

Typeset by LaserScript Limited, Mitcham, Surrey
Printed and bound in Great Britain by
Biddles Ltd, Guildford and King's Lynn

British Library Cataloguing in Publication Data

Dewar, David
    Urban markets : developing informal retailing.
    1. Urban regions. Unemployment
    I. Title   II. Watson, Vanessa, *1950–*
    331.13'791732

    ISBN 0-415-03813-8

Library of Congress Cataloging in Publication Data

Dewar, David, B.A.
    Urban markets : developing informal retailing / David Dewar and Vanessa Watson.
        p.   cm.
    Bibliography: p.
    Includes index.
    ISBN 0-415-03813-8
    1. Markets — Management.   2. Informal sector (Economics)   3. Retail trade
    4. Urban economics   I. Watson, Vanessa.   II. Title.
    HF5470.D45   1989
    658.8'7 — dc20
                                                                89–32095
                                                                CIP

Contents

List of plates

List of figures

Preface

This work arose from a concern with growing levels of poverty and unemployment in less-developed countries and from an increasing recognition that, in them, very small-scale, self-generated, economic activity will have to provide an important means of survival for the very poor. Exposure to, and work in, a wide variety of low-income urban contexts by the authors gave rise, over time, to a number of beliefs. First, the facilitation of urban markets – defined as the physical agglomeration of small traders and producers – is a potentially powerful policy instrument that can be used to facilitate the efforts of the urban poor to generate income. Second, while markets sometimes occur spontaneously, management intervention is frequently required to reduce clashes and conflicts: the more dense and complex the urban context, the more this may be necessary. Further, many spontaneously formed markets reveal, on closer examination, a high degree of internal organization. Frequently, too, the form of this organization is exploitative, in the sense that it favours relatively larger traders at the expense of smaller, more fragile ones. Third, the success of markets is profoundly affected by the way in which they are located, structured, and administered. There is clearly a considerable need to think carefully about market systems and their management. Despite this, the authors were unable to find any literature dealing with these issues: this book is an attempt to fill that gap.

Observation of markets in many places, particularly Africa, the Far and Middle East, and Latin America, gave rise to the conviction that there were several common issues which affected the success of markets. To test this, a specific research project into the operation of markets in seven cities was initiated. The study confirmed this belief. This book draws upon this collective experience to produce guidelines for the location, design and management of markets. It is intended as a practical guide to decision-makers and to urban planners and designers, both within the public sector and in private practice. The empirical evidence used to illustrate points made is primarily derived from the investigation of the seven cities, although other cases are used where appropriate.

Having stated what the book is intended to accomplish, it is equally important to stress what it does not attempt to do. First, it does not attempt to promote, in any way, the formalization or 'neatening' of informal economic activity. The basic philosophy underpinning the argument here is one of minimalist intervention: the most appropriate level of intervention is the minimum necessary to ensure market success and to allow opportunities to trade to as many people as possible. Under this umbrella, intervention should seek to give traders maximum possible room to manœuvre. However, the approach advocated recognizes that conflicts may occur between the desires of traders and the legitimate concerns of other urban interest-groups. Sensitive intervention seeks to resolve these clashes and thus to ensure that small-scale economic activity is promoted to the greatest possible degree. In essence, therefore, the approach does not argue for the formalization of the informal sector, but for the legalization of activities frequently regarded as illegal, and for the provision of market spaces as essential forms of urban infrastructure.

Second, it is a fundamental belief of the authors that all planning action must be defined and shaped by the context in which it occurs. The book does not, therefore, attempt to provide a 'blueprint' for the planning of markets. Rather, it establishes a set of principles to guide decision-making on market establishment and to sensitize decision-makers about some of the issues involved in a markets policy.

Third, the book does not purport to represent a comprehensive account of the entire market system of the cities studied, or, indeed, even of the markets which are depicted. The markets chosen are not important in their own right. They are used simply to the degree that they illustrate some of the issues we wish to highlight.

Sincere thanks are extended to the Chairman's Fund of Anglo-American and De Beers, and to Metboard, for financing the research, and to the Chairman's Fund of Anglo-American for subventing publication in order to allow the use of photographs. Diagrams and maps were produced by Farouk Stemmet and Duncan Rendall, typing was cheerfully undertaken by Suzette du Toit, Merry Dewar designed the index, and final editing and proof-reading was done by Kathy Forbes; to all, we are grateful.

Chapter 1

Introduction: Urban markets and informal-sector stimulation

Considerable rhetorical emphasis is being placed, in many countries, on the potential role of the informal sector in alleviating poverty and unemployment, and there is a growing call for the 'stimulation' of this kind of activity. There is little clarity, however, about precisely *how* such stimulation should occur. This book focuses on one potential instrument of stimulation: urban markets which cater for small-scale vendors. Specifically, it seeks to provide insights into issues relating to the role, location, physical structure, financing, and administration of such markets. Its origins stem from research initiated in 1980 (Dewar and Watson 1981) into the question of the urban informal sector and its stimulation. In the course of that work it became apparent that the stimulation and facilitation of markets to accommodate small-scale traders and manufacturers represents one significant means of promoting small-scale economic activity in urban areas. It also indicated that the efficacy of markets, and thus their effectiveness as a policy tool, is strongly informed by their location, structure, form, and administration.

To understand the significance of the issue, and the concerns underlying the observations which follow in this book, it is necessary to locate the question of markets in the broader context of informal-sector stimulation.

Informal-sector policies

Over the past few decades, attitudes of policy-makers towards the informal sector have undergone some major changes (Sanyal 1988). Development theorists and policy-makers alike in the 1940s and 1950s regarded the informal sector as the declining remnant of pre-capitalist economies. As such, it was perceived to be an inefficient, backward, irrational, and frequently unhygienic form of economic activity. It was assumed, in keeping with general tenets of the then dominant modernization paradigm, that as the less-developed economies progressed from a 'traditional' to a 'modern' state, the informal sector would be incorporated into the 'modern sector' of the economy and would gradually disappear. In policy terms, there were attempts in many parts of the world, at both central and local levels of government, to restrict the operation of the informal sector and, in particular, to exclude it from residential areas and the main commercial centres of cities.

In the late 1960s and 1970s, both this perception and its associated policy approaches underwent a significant reversal in many countries. In less-developed countries particularly, problems such as poverty, inequality, and unemployment were manifesting themselves in political dissatisfaction, and policy-makers were concerned about finding alternative ways to ameliorate such problems. It was becoming clear, too, that previously accepted, indeed almost unquestioned, approaches to economic development, which assumed that the benefits of economic growth would 'trickle down' to the lower income groups more or less automatically, were not always resulting in widespread improvements in welfare levels, particularly amongst the poor. Strategies which consciously promoted a wider distribution of economic benefits and, particularly, greater employment, were required.

A watershed event in this regard was the launching, in 1969, of the International Labour Organization's (ILO) 'World Employment Programme'. While acknowledging the need for economic growth, the agency argued that 'part of the difficulty is structural, in the sense that many of these difficulties will not be cured simply by accelerating the rate of growth' (International Labour Office 1972: xi). Accordingly, it initiated a series of country-wide studies to evolve employment-orientated strategies of development.

Perhaps the best-known and most influential of these was the Kenya Mission Report (International Labour Office 1972), which advocated an economic-development strategy for that country termed 'redistribution with growth'. The report identified as a central issue the need to stimulate local demand in order to lay the basis for a broader, less import-dependent economic system. To achieve this, the report placed emphasis on the redistribution of income, the stimulation of rural and agricultural development, the reorientation of industry towards local raw materials and demands, the promotion of labour-intensive practices, the use of appropriate technology, and increased economic self-reliance.

Most significantly for this discussion, however, it identified the informal sector as having the potential to play an important role both in providing employment and in contributing to a reorientation of the economy. As opposed to previous conceptions of the informal sector as a stagnant and undesirable phenomenon, the Kenya Report emphasized that the informal sector, which provided employment for between 28 and 33 per cent of the working population of Nairobi, was by no means unproductive or declining. Instead, the authors argued, it was competitive, labour-intensive, made use of locally produced materials, developed its own technology and skills, and businesses were usually locally owned. According to the report, therefore, it not only played an important part in reducing Kenya's employment problem, but also helped to serve the lower end of the consumer market without making excessive demands on foreign exchange or imported capital goods. In policy terms, the report recommended the direct encouragement by the government of the informal sector as a partial solution to the problems of unemployment and poverty.

This switch in attitude towards the informal sector was by no means universally accepted in policy terms. In many countries, the perception of it as an aberration and a nuisance has continued. However, the ILO study did have the effect of focusing attention on the phenomenon, and it unleashed, in subsequent years, numerous empirical and theoretical studies on it.

While it is not the intention here to undertake a review of theoretical developments in the field of informal-sector studies, these studies have raised a number of debates and issues which are directly relevant to policy formation in this field. Since this book is about a dimension of policy formation, it is important that the authors' personal positions on these debates are clarified for the reader from the outset.

Definitional debates

The first debate centres on definition. Over time, many and varied attempts have been made to define the phenomenon of the informal sector, and it is clear that 'an analytically verifiable definition of the informal sector still remains to be constructed' (Sanyal 1988: 66). It has, however, become apparent that the definition will properly vary according to the *purpose* for which, or the philosophical position from which, the definition is being made. One of the central issues which has emerged in the definitional debate is whether the focus of attention should be on the *business* or *activity*, or on the *household unit*. The approach taken here is that, from a theoretical perspective, and in terms of explaining the nature and characteristics of these activities, the household unit is clearly a crucial variable in terms of both income generation and income distribution. However, if the concern is with policy and the impact of outside stimulatory agents and instruments on informal activities, then properly the emphasis is better placed on the economic enterprise.

Further, it is misleading, from a policy perspective, artificially to separate informal, small-scale economic activities from larger, more formal ones. They do not operate in separate economic circuits: indeed, they are vitally inter-related. They also have similar economic requirements and respond to similar stimulatory or depressive impulses, although the form of these may vary. From a policy perspective, therefore, the term 'informal sector' simply focuses attention on economic enterprises at the bottom end of a continuum ranging from very small to very large businesses. It is used here as a convenience, since the widespread and popular use of the term 'informal' conjures up an image of less stable, more oppressed and fragile, and sometimes impermanent economic activities, and these represent the focus of the activities which are central to this work (Plate 1).

Plate 1 Informal selling: Crossroads, Cape Town.

The informal sector is usually viewed as a fragile and impermanent economic activity.

Debates relating to linkages

The second debate stems from a recognition of the numerous and complex linkages which exist between the 'formal' and 'informal' sectors, and hinges on two questions: first, are the linkages of a benign or exploitative nature; and second, is the lower end of the economic continuum capable of responding to promotional policies or is it not – is it *evolutionary* or is it *involutionary*? (Tokman 1978).

Basically the debate has revolved around the question of how capital accumulation takes place in the informal sector. On one hand, it is argued that these activities do generate surpluses unless unduly repressed by the law: in this case it is assumed that the linkages are benign (Weeks 1975, Sethuraman 1981). On the other hand, it has been argued that the informal sector is incapable of significant accumulation because its level of surplus is dictated by the accumulation process in the formal sector: that is, the linkages are exploitative (Bromley and Gerry 1979, Moser 1984).

Those who argue in favour of a benign relationship emphasize the crucial role played by the informal sector in the circulation process by being located near customers, by providing credit, by selling in units as required, and by targeting products specifically at the needs of the low-income market (Tokman 1978). The informal sector, therefore, generally serves a specifically poor market and in this sense remains complementary to the formal sector (Plate 2). Its capacity for accumulation can then be enhanced by its access to the expanding markets in the rest of the economy: its growth may be 'evolutionary'.

Plate 2 Making sjamboks (whips): Crossroads, Cape Town

The informal sector frequently targets products specifically designed for the low-income market. These sjamboks (whips) made in Crossroads, Cape Town, serve as symbols of authority as well as protective instruments in low-income areas.

Those who argue that the relationship between the formal and informal sector is exploitative have generally taken as their point of departure the theory of unequal exchange as fundamental to an explanation of regional inequality. In brief:

> The process of accumulation in the developed countries assumes the characteristic that productivity gains are retained within the centres, while simultaneously the gains in productivity registered in the periphery are appropriated through different mechanisms ... such as ... international price determination and market control ... and institutional arrangements fostered by transnational capital.
>
> (Tokman 1978: 1067).

The relationship between the formal and informal sectors is analyzed as a subcomponent of this process, whereby the economic surplus generated in the informal sector is transferred to the formal sector. There are two major mechanisms through which surpluses may be transferred from the informal to the formal sector.

First, the informal sector lacks access to basic resources of production because these resources are monopolized by the formal sector. Thus

the oligopolistic organization of the product markets leaves for informal activities those segments of the economy where minimum size or stability conditions are not attractive for oligopolistic firms to ensure the realisation of economies of scale and to guarantee an adequate capital utilization.

(ibid.: 1069).

Second, the informal sector is forced into a position whereby it must pay higher prices for its purchases, but can only ask lower prices for its outputs, the difference being reaped by the formal sector. Prices of purchases are generally higher because small operators can only buy small quantities and they do not have access to credit facilities, while prices for their products, mostly services, are lower because of the market they depend upon (ibid.). Because the ability of the informal sector to accumulate is thus limited by its relationship to the formal sector, the ability of the informal sector to grow or respond to promotional policies is limited, and the sector as a whole is said to be 'involuting'.

The question of interconnectivity between small businesses and large, and the implications of this for the evolution or involution of small businesses, is thus an important one. These authors believe that the weight of evidence favours the position taken by Tokman, who questions the simplicity of both the 'benign relationship' and 'exploitative relationship' schools. The informal sector comprises a great variety and range of enterprises and it is not possible to generalize as to their relationship with the formal sector. 'The problem', Tokman says, 'is to determine how strong the subordination is, and whether there is room left for evolutionary growth.' (ibid.: 1071). While both prices and markets may indeed be determined outside the informal sector, there are also numerous ways in which the informal sector can maintain a share of the market. Tokman (ibid.: 1073) quotes, in the case of small retail outlets: 'Location, owner–customer personal relationships, credit, infinite possibilities of product sub-division, permanent presence because of the non-existent "business hours", etc...'. It is evident that in any study of small businesses, a careful disaggregation needs to take place, to determine exactly where the possibilities of expansion or contraction are likely to be.

Nevertheless, there is no *intrinsic* reason for rejecting the introduction of policies to stimulate this kind of activity. Certainly, it is untenable to suggest, in contexts characterized by high unemployment and an increasing competitive imbalance in favour of larger, 'formal' businesses, that the repression of informal sector activities should continue. It can be assumed that if a more facilitative policy towards the informal sector were adopted, it would be possible for a greater number of people to improve their survival chances, or at least to supplement their income, in this way. Similarly, it is sensible to introduce policies which seek to strengthen the relative position of economic concerns at the lowest end of the continuum of economic activity, in a proscriptive and not a prescriptive way: policies should seek to create new and better opportunities and to remove the major regulatory and other obstacles under which small disadvantaged enterprises have to operate.

Debates relating to 'solution' or 'survival' strategies

The third debate is whether or not the informal sector represents a potential solution to unemployment in developing contexts or whether it should properly be viewed as a survival strategy on the part of the urban poor. The present authors believe that the central issue, in developmental terms, is that the problem is not employment creation *per se*, but improvement of the quality of life of the urban poor. As Weeks (1973) argues, the problem in most developing countries is not only one of people enduring involuntary idleness but that 'the vast majority of urban dwellers are working long hours in a debilitating climate, while burdened by energy-draining diseases and parasites, to earn a marginal income.'

While issues of poverty and employment are obviously interrelated, the distinction drawn recognizes that it is possible to increase employment without necessarily alleviating problems of poverty and inequality in the long term. In this regard, the overwhelming weight of evidence indicates that, in a great many informal enterprises, profits are low, growth is slow or non-existent, working hours are long, working conditions are highly insecure, and wages are well below the average paid in the formal sector (Bromley and Gerry 1979, Dewar and Watson 1981).

Further, the close interconnections with the formal sector do place restrictions on the aggregate scale of accumulation in informal-sector concerns, although not necessarily in every case. Clearly, then, the informal sector offers no universal panacea to problems of poverty and material deprivation: in policy terms it is vital to generate maximum job creation and remuneration in the formal sector. To the degree that policies aimed at the informal sector divert attention away from a holistic interpretation of the problem or create a sense that the problem is 'under control', they may in fact be dangerous. Nevertheless, informal-sector activities comprise, for many of the urban poor, the only, or, at least a vital, supplementary source of income, and are critical to their survival. There clearly is scope, therefore, for policies aimed at increasing income-generating opportunities for these types of operators.

A particular advantage of strategies aimed at stimulating the bottom end of the economic continuum is that they not only contribute to employment creation but constitute a direct attack on poverty. The people engaged in these concerns are usually (though not exclusively) drawn from the poorest strata of society. Stimulatory policies are thus focused directly upon the poor and, therefore, have the best chance of reaching that group. Additionally, the smallest, most fragile concerns, be they in the commercial, manufacturing, or service sectors, primarily serve the poorest groups. Consequently, the goods and services produced and distributed are mainly in response to the specific needs of the local, poorer, market (food, clothing, shelter, hygiene, and so on).

Finally, policies aimed at strengthening these small concerns tend to lead to greater re-circulation (or second-round circulation) of capital amongst the poor. Obviously, there is considerable leakage: nevertheless, the poor potentially benefit more than from policies aimed simply at stimulating the growth of larger concerns.

Some policy implications of research

It seems, therefore, that there is a case to be made for introducing stimulatory policies to facilitate activity in the informal sector, but great care must be taken as to how this is done. Research into the operation of the informal sector suggests a number of specific guidelines in this regard.

First, the economy is highly complex: it is characterized by fine-grained networks of interlinkage, inter- and intra-sectorally, between large and small firms, between formal and informal concerns, and so on. The relationships between elements and actors in these situations are neither equal nor necessarily benign: stronger, more powerful elements tend to dominate. Two interrelated implications flow from this. One is that 'filtration' policies (policies directed at relatively stronger businesses or sectors, in the hope that benefits will filter down to weaker, more fragile, enterprises) are almost certainly doomed to failure. If policies are to benefit the smallest, most fragile economic enterprises, they must be consciously formulated in such a way that these enterprises can benefit: policies must be targeted at them.

The other implication is that the scope for exploitation is great and the more fragile the economic enterprise, the more susceptible it is to exploitation. Consequently, there is always a tendency for the benefits of even those policies directed at the weaker elements to be appropriated by larger, more powerful concerns, unless specific efforts are made to overcome this.

Second, empirical work internationally indicates that it is incorrect to view the informal sector either as static or homogeneous (Moser 1978). From a policy perspective, one of the most important distinctions relates to the motives of people operating in the informal sector: these are complex and vary considerably over time and space. At any one time, certain aspects of informal-sector activity may be expanding while other aspects are in decline. Thus, for example, in a study in Dakar, Le Brun and Gerry (1975) distinguish between 'petty producers' who saw the accumulation of wealth as the main aim of their activities, and 'artisans and traditional petty producers' who simply aimed to generate a subsistence income.

The difference between these two categories was also reflected in their area of operation: petty producers were more often producers and sellers of foodstuffs or general consumer goods; artisans and traditional petty producers were more concerned with making articles, often to fill client orders, and were facing considerable competition from mass-produced consumer goods.

The importance of this distinction between operators who are genuinely entrepreneurial (in that their central motivation is economic growth and accumulation and that they re-invest profits to these ends) and those who are 'survival' operators (in the sense that short-term profits are frequently consumed) is not only academic. Frequently, operators with different motives adopt different temporal rhythms of operation: some operate full-time; others over certain hours of the day only; others even on a limited number of days of the year. If policy is to be successful, it must facilitate the operation of all these types of concerns. The danger is that it selectively benefits full-time (and usually economically stronger) enterprises at the expense of more impermanent or intermittent operations.

Third, the implication of much of the previous discussion is that the most effective policy actions are those which increase the room to manœuvre for small operators by creating additional and better options, rather than those which attempt to identify and prescribe particular modes of behaviour, activity, or action as being 'best', or which are too finely targeted at specific sectors or activities.

The reason for this lies in the complex nature of urban economies. They are so complex, in fact, that their precise forms, interlinkages, and interrelationships (given available time-frames for research and the rate of change within the economy itself) usually defy the fine-scaled and comprehensive analysis which is needed to allow for fine-tuned targeting. Further, changes in interrelationships occur rapidly and in a frequently unpredictable form. In this type of situation, the more specific the policy action, the greater the tendency for it to be appropriated by relatively stronger elements and enterprises at the expense of weaker, more fragile ones. The very introduction of a sectorally targeted policy measure frequently causes a rapid restructuring within the economic sector affected, with stronger elements seeking, more effectively and aggressively, to capitalize on the externally induced advantage and, in so doing, frequently displacing more fragile enterprises. This tendency towards appropriation and distortion is, of course, strongest in relations between the formal and the informal or petty-commodity sectors. As Moser (1978: 1062) argues, 'because of the dependent relationship between the capitalist sector and the petty commodity sector, policy solutions designed to assist the latter almost invariably end up by promoting the former.' However, it also occurs within the informal sector itself.

The implication of the argument is that in capitalist systems, it is impossible to pre-empt the operation of the market. The best that can be done is to create the widest possible range of opportunities and to make every effort to ensure that even the most fragile concerns can benefit from them. The wider the range of choices, the greater the 'manœuvring' or 'survival' space for small operators and the greater the possibilities for all.

Actual policy responses

Subsequent to the articulation of the redistribution-with-growth approach, two diametrically opposed approaches to development have emerged. One calls for the pursuit of accelerated economic growth, via the functional integration of economies into a world economic system. The benefits of this growth, it is argued, will automatically penetrate throughout the fabric of society. The other argues that development will not automatically result from the functional integration of economies. Indeed, it can only be pursued in relation to a defined territorial base, and territorial (as opposed to functional) integration is required. In particular, it calls for direct attacks on poverty via policies which seek to mobilize local resources towards the satisfaction of the basic needs of the poor.

Both approaches emphasize the role of the informal sector in promoting employment. In practice, development planners have primarily advocated policies to reduce *constraints* on informal-sector activity (Sanyal 1988). These have been of two kinds. On the one hand, there are those aimed at reducing *internal* constraints on informal-sector growth (such as lack of education of informal-sector operators, poor management-skills, lack of capital, and use of low-productivity technologies). On the other hand, there are those aimed at reducing *external* constraints (such as low exchange rates which favour formal-sector production, government licensing requirements which set unrealistic standards, government purchase of supplies from exclusively formal-sector enterprises, and lack of access on the part of the informal sector to capital from formal channels).

Available evidence indicates that these policies have achieved limited results. Clearly, however, this is not enough: more pro-active policies are required. The current decade has been 'marked by negligible or negative [economic] growth and increasing poverty' (ibid.: 74). The World Bank has predicted low rates of increase in urban formal-sector employment, while the urban labour force in developing countries will continue to grow at current high rates. Further, International Monetary Fund austerity measures have reduced government spending and made the pursuit of traditional economic promotional policies particularly difficult.

The informal sector in most developing countries, therefore, is likely to become even more important. Sanyal is correct when he states that for

> the large and increasing number of families below the subsistence income, a set of mutually complementary policies, related to shelter and service provision, income generation and, possibly, subsistence food production, is required. This in turn will require a fresh look at the spatial structure of the urban economy.
>
> (ibid.: 80).

The focus of this study: urban markets

This book is concerned with just one aspect of urban economic structure: urban markets for very small-scale economic activities. The facilitation of a system of urban markets – defined as the physical agglomeration of small traders and producers – is a potentially powerful instrument for stimulating informal-sector activity. If appropriately handled, it increases the sense of security of operators (an essential condition for rational economic behaviour) by 'legalizing' their activities, and it allows more small entrepreneurs in genuinely viable locations than would be the case if such a policy did not exist. Further, it improves the trading environment for small traders at little or no overhead cost, thereby increasing their competitive position *vis-à-vis* larger, more formal enterprises. Preliminary research by the authors in a wide variety of contexts indicated, however, that the physical, financial, and administrative design of markets has a fundamental and far-reaching impact on the kinds and numbers of operators who establish themselves in a market, and the ability of these operators to survive economically.

This book seeks to provide insights into these issues. Its focus is on cities in less-developed countries, but this does not imply that markets do not exist, or are unimportant, in urban environments in developed countries. Arguably, however, the magnitude of informal-sector activity is much greater in less-developed countries, and there is a greater focus on basic (as opposed to luxury or up-market) products and services.

Approach

In order to test initial impressions and to understand further the factors affecting the success of markets, seven cities in seven different countries were chosen for more intensive study. All have a long-established tradition of planned or spontaneous markets, catering to the needs of small vendors. Most of these countries, too, have instituted significant policy changes over time. It was felt, therefore, that a great deal could be learnt from them with respect to factors that hinder or promote small-scale vending and, in particular, that affect the performance of markets. The cities chosen were Bombay (India), Colombo (Sri Lanka), Bangkok (Thailand), Hong Kong, Singapore, Taipei (Taiwan), and Harare (Zimbabwe). In all, some sixty-six markets were documented and analysed. Markets in other cities (such as Bulawayo, Zimbabwe; Blantyre and Llongwe in Malawi; and Durban in South Africa) were visited. The field-work was not as extensive in these cases, and the seven named provide most of the empirical information contained in this document. The supplementary visits did, however, confirm the general observations articulated here.

It is important to emphasize that the study does not purport to represent a comprehensive record of market activity in each of the cities. Indeed, the cities in their own right are unimportant to the purpose of the study, which is to uncover significant lessons relating to the design and management of markets. The cases merely provide empirical laboratories through which, either by omission or commission, issues are raised: where appropriate or useful, examples experienced in other cities are used. Analysis in the sixty-six cases was carried out through physical observation and measurement as well as through interviews with market administrators and those involved with market policy.

Typology of markets

Descriptively, five main categories of differentiation exist in relation to urban markets.

Nature of supply

The main distinction here is generally between *wholesale markets*, which primarily distribute in bulk and which are the principal source of supply for retail markets and other forms of outlet; and *retail markets*, which break bulk and deal in much smaller quantities and which directly serve the consuming public. The focus of this study is upon retail markets, although reference is made to the system of wholesaling in relation to the retail market system.

Function

Generally, even though many markets are mixed and accommodate a wide variety of products and services, they usually reflect a functional emphasis (for example, food, clothing, household goods, and so on).

Plate 3 Informal street market with most infrastructure provided by sellers

Degree of formality

A wide range of options exists about the degree to which externally provided shelter, infrastructure, and services are provided in markets. At one end of the continuum, nothing is externally provided (as, for example, some informal street markets [Plate 3]); at the other, almost everything is externally provided (this may take the form, for example, of a market building with a full range of infrastructure [Plate 4]).

Plate 4 A formal market-building with high levels of infrastructure: Hong Kong

Form

Urban markets take many different physical forms. At the most basic level, the primary distinction is between linear markets, the form of which is usually directly informed by lines or channels of movement and nucleated markets, which take more concentrated form at particular *points* within the city.

Time of operation

Finally, markets vary in terms of their time of operation. The primary distinction here is between permanent and temporary markets, although a range exists in terms of the pattern of periodicity (for example, night markets, morning markets, weekend markets, and so on). Significantly, too, different traders frequently adopt different temporal rhythms within any one market.

Clearly, the appropriateness of market type will vary with physical, economic, political, and social context: no one type is inherently better or worse than any others and the initial selection of type depends on context-specific analyses. Importantly, however, the development of markets needs to be understood processionally and categories of differentiation may not be mutually exclusive when viewed over time. Thus, for example, the functional emphasis may change with time; informal markets may accumulate infrastructure, shelter, and so on progressively, and become more formal over time; linear and nucleated forms are not necessarily mutually exclusive – the one is frequently an emphasis in a progression in which both forms are present; and temporary markets may become permanent or vice versa. An important part of the skill in the design and administration of a markets system is in recognizing and allowing for the prospect of change and identifying the factors that may generate this.

The comments which follow, therefore, are not tied to one or another category or combination of categories. Rather, an attempt has been made to pitch them at a level of generality which extends beyond the differences. Where reference is made to certain types only, this is clearly indicated.

Structure of the study

The study is structured in the following way. Chapter 2 outlines the major conclusions of the analysis: it constitutes a primer on the design, administration, and financing of urban markets. Chapters 3 and 4 provide case material from the analysis to underscore the conclusions in Chapter 2. Chapter 3 deals with the nature of informal selling, with attitudes towards, and approaches to, management of informal vending in general and urban markets in particular, and with the wholesaling systems, in the cities mentioned. Chapter 4 presents case material on the location, layout, and infrastructure of actual markets in these cities. To enable ease of reference, this chapter is structured under the same headings as Chapter 2. Obviously, no attempt is made to cover all the markets analysed and the presentation of material from any market does not reflect a judgement on the relative success or failure of the case in question. Material has been chosen simply in terms of the degree to which it illustrates or illuminates the issues raised.

Urban markets: some issues relating to their location, design and administration

This chapter attempts to articulate some of the insights gained from the analysis of markets in the seven cities identified. It is important to emphasize from the outset that the comments which follow are not advanced as rules or tenets which should be adhered to on all occasions. Indeed, an overwhelming lesson which emerges from the analysis is that there are many more ways of achieving successful performance than can be laid down in a given set of rules. Some of the most successful markets analysed reflect relatively spontaneous responses on the part of traders to situations where there is a shortage of space or awkwardly provided space: that is, where there is some form of contextual constraint. There is clearly always room for ingenuity and creativity and the design and implementation of any market must always be based on contextually specific analyses. Nevertheless, it is believed that the *issues* and *principles* identified here do reflect the main ones which need to be considered in the design and management of a system of urban markets, or in the design of any particular market. We believe that much of what follows is little more than common sense. The number of failures observed in the analysis, however, convinces us that there is a considerable need for people involved in market design and administration to be reminded of basic tenets. This chapter is advanced in that spirit.

The chapter is structured under the following heads: attitudinal issues; location; use mix; physical layout; market infrastructure; supply-side support (with particular reference to wholesaling); and administration and management.

Attitudinal issues

This section argues the case for implementing an urban-markets policy to facilitate the operation of very small traders, and clarifies the attitude which should underpin that policy. This question can only be examined in relation to the broader issue of defining an appropriate attitude to the informal sector in general and to hawking and vending in particular. In practice, three attitudes to it are most commonly observed.

Attitudes to the informal sector

The first is that informal vending should be controlled to the greatest possible degree. The second (represented, for example, in Hong Kong, Singapore, and Harare) is that it is a social evil and should be phased out over time. The third is that a *laissez-faire* attitude should be adopted towards it. This approach is represented, *de facto*, in the case of Bombay, although rhetorically it is supposed to be controlled. Although periodic attempts are made at control, via police raids, in practice hawking and informal vending continue unabated in an almost entirely non-regulated way.

Implications for an urban-markets policy

These different attitudes have profoundly different implications for how urban-markets policies are viewed. In the first case, no positive policy can exist: the policy emphasis is entirely upon regulatory control. In the second case, the emphasis is upon using markets to 'tidy up' and 'formalize' informal vending and to remove it from the streets. In the third case, again no formal policy can exist, although certain reactions – such as street cleaning – may be instituted to accommodate it.

None of these attitudes or associated policy responses, however, is really appropriate to conditions in developing countries. The overriding contextual realities in these countries are very high levels of poverty and unemployment. Further, it is impossible in these cases to have confidence that all potentially economically active people who are seeking urban employment will be absorbed in the formal economic sector of cities within the foreseeable future or that poverty will be eradicated in the short to medium terms. In these circumstances, small-scale, self-generated employment will *inevitably* have to play a greater role as people engage in survival strategies. Even though it provides no ultimate solution to poverty and unemployment, therefore, every possible opportunity for the generation of small-scale, informal economic activity should be capitalized upon.

The most appropriate attitude to informal vending, it is argued here (and almost no examples of it can be found in practice), is one which views it in a positive light but which seeks to accommodate it in a way that resolves, to the greatest possible degree, potential conflicts between traders and other urban interest-groups which are often the cause of the introduction of repressive actions against traders. In terms of this attitude, markets should be seen as potential instruments of economic stimulation. The success of a markets policy, therefore, should not be measured only in terms of how well it accommodates existing informal trading but also in terms of how it *increases* opportunities for greater numbers of traders, how it promotes this economic growth and expansion, and how it accommodates the needs of even the most vulnerable operators. A markets policy, therefore, is not seen just as a reactive 'cleaning up' mechanism, but as a pro-active instrument of stimulation.

The value of a positive markets policy

The analysis conducted on market systems in various contexts reveals that there are several reasons why a positive markets policy is a potentially powerful instrument of assistance to informal operators.

First, a large number (arguably, the majority) of small-scale informal-sector operators in the cities of developing countries are involved in the retailing sector. Potentially, therefore, investment in market infrastructure which, it is argued here, should be an essential form of public investment in all cities, is capable of benefiting very large numbers of these small operators. Further, it can do so without discriminating against those who cannot prove a past record of operation or of profitability.

Second, the stimulation of markets represents almost the only way in which very small operators can gain access to central, viable locations in the city. One of the biggest problems which very small operators face is that of physical marginalization. Because of their very low rent-paying capacity, they tend to be forced into the peripheral areas of the city, where the potential for profit generation is slight (Plate 5). By capitalizing upon the *collective* potentials of individual traders, the provision of markets increases their capacity to capture central, more viable, locations.

Plate 5 Frequently, small operators are spatially marginalized in that they are forced to operate in low-density residential areas where the local market is thin: Cape Town.

Third, the physical concentration of large numbers of traders increases their drawing capacity. The potential of very small-scale traders, operating individually to attract customers, is extremely limited. The physical agglomeration of large numbers of small traders in markets, however, and the potential for comparative buying which this sets up, in aggregate provides a magnet capable of competing with much larger commercial establishments.

Fourth, the physical proximity of a number of traders establishes the potential for other forms of mutually advantageous co-operation. Thus, for example, cheaper buying may become possible, either because the demand of a number of traders may make delivery of bulk supplies by wholesalers feasible; the collective use of vehicles may make wholesale markets accessible to individual traders who would otherwise effectively be denied that access; and so on.

Fifth, markets in low-income areas can provide an important service to consumers. One characteristic of low-income communities is that, in marketing terms, many people are trapped in very localized areas, since they are unable to afford the costs associated with overcoming the friction of distance between them and regional or subregional facilities (for example, transport costs are frequently too high, relative to the cost of purchases, to make longer journeys feasible; they are unable, for reasons of child-minding, to move far from the house; they are unable to afford refrigeration or bulk storage facilities necessary to enable bulk buying; and so on). However, the thresholds contained within these local areas are not large enough to support competition between larger, formal suppliers. Accordingly, suppliers located in these areas frequently find themselves in quasi-monopolistic conditions and can charge exorbitant prices. When informal-market facilities exist, however, intense competition may occur between many, far smaller, operators and the consumer benefits both from lower prices and from the potential for comparative buying. Additionally, if properly designed, these markets have the potential to become the social and recreational centres of the local communities.

Sixth, from the perspective of urban management, a positive markets policy contributes to the resolution of potential conflicts between the need to assist small-scale vending, on the one hand, and the need to satisfy demands of other urban interest-groups, on the other. Four complaints are commonly made about informal vending: frequently these underpin the repressive or restrictive policies employed in relation to informal-sector operators in many cities of the world.

One is that informal traders impair vehicular or pedestrian flows and thus contribute to congestion and urban inefficiency. A second is that this form of activity is unhygienic and constitutes a threat to urban health. There are frequently three heads to this argument: facilities for the hygienic storage and preparation of produce (for example, the washing of fruit and vegetables) do not exist; the regular inspection and monitoring of hygiene levels is not possible; and toilet facilities cannot be provided. The weight given to these arguments relating to traffic and health is reflected in the fact that control of informal-sector activity frequently resides in the traffic and health departments of urban authorities. A third complaint is that it is a major cause of litter and environmental degradation. Finally, it is held that informal vending is a form of unfair competition to 'formal' traders: informal vendors have limited overheads and thus can undercut prices; they locate in interceptor positions between pedestrian flows and formal shops, thereby 'stealing' custom; and they consciously or unconsciously seal the entrances to shops, thereby making access to them extremely difficult.

Plate 6 Informal traders frequently place great emphasis on cleanliness and presentation: Bangkok

It is important to acknowledge that these arguments *may* have some foundation: the almost anarchic operation of informal-sector activity in Bombay, for example, unquestionably creates very severe problems. It is equally important to stress, however, that such arguments are not *intrinsically* correct and that they are frequently overstated. It depends on the scale at which, and the way in which, the activity is conducted. Thus, for example, if properly managed, informal vending need not increase traffic congestion significantly, or negatively affect pedestrian circulation; competitive economic behaviour may well resolve the issue of hygiene, in the sense that it is poor business practice to sell produce which is unhygienic or inferior in any way. The enormous emphasis placed on product presentation and cleanliness (Plate 6) in many of the markets analysed, for example, and the extremely selective nature of consumer purchasing in low-income communities, bears testimony to this: frequently, places of informal vending are kept extremely clean by operators and are made environmentally delightful, again because the environmental condition affects business conditions; and the amicable and frequently mutually beneficial relationship between formal and informal operators found in countless places refutes the characterization of inevitable conflict.

The central point, for the purposes of this discussion, however, is that the stimulation of market activity at carefully selected locations allows potential conflicts between legitimate concerns and interests to be resolved to an acceptable level: markets, and particularly market systems, can be designed in such a way that traffic problems are reduced; facilities to promote hygiene can be provided if the scale of activity warrants this; co-ordinated garbage collection and cleaning operations can be organized; and conflicts between formal and informal traders can be minimized - through sensitive physical design, which uses volumes of space to provide protection, through negotiation (for example, over time of use of an area or the creation of specialized areas) and through monitoring and control.

Plate 7 Samsen Road market: Bangkok

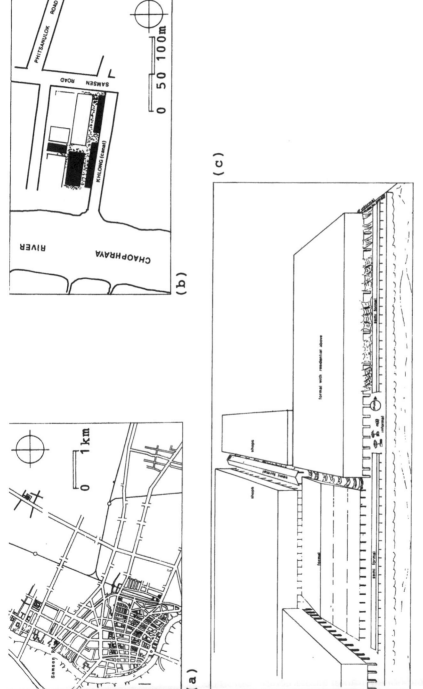

Figure 1 Samsen Road market, Bangkok: (a) Bangkok; (b) local area map; (c) Samsen Road market

An example of a market which provides trading opportunities for large and small traders. There are two formal market buildings: these are on the ground floor of buildings with residences located above them. There are two formal market buildings located above them. There are two long sections of corrugated iron roofing which give shelter to informal traders (see Plate 7). Canvas has been strung between the roof and building to extend the shelter. Finally, the pavement on Samsen Road at the entrance to the market provides a location for some of the very smallest traders who are in direct proximity to passing pedestrians and traffic flows.

Significantly, conflict resolution will frequently demand compromise: the situation may not be ideal in terms of traffic circulation, hygiene, or from the perspective of either informal or formal traders. Provided a facilitative attitude and not a control-obsessed one dominates policy direction, and provided there is a realistic acceptance of the possibility of conflict which needs resolution, however, the system overall will work better than one which is driven by the desire to satisfy all the demands of one or another set of concerns or interest groups.

The importance of basing informal-sector policy on an attitude of facilitation and not control has significant implications for how a markets policy, which properly should comprise just one subsector of such a stimulatory policy, should be viewed. Appropriately, the promotion of markets should not be seen as a substitute mechanism for 'neatening up' hawking. What is required is a physical *system of marketing infrastructure* which opens up the widest possible range of trading opportunities. Inevitably, such a system should have a strong hierarchical dimension: it should make provision for a range of trading opportunities catering for very large traders as well as very small ones. The point to emphasize is that in such a system the creation of spaces for the smallest traders is as important as space creation for the larger ones. If the markets policy in practice *reduces* the range of marketing opportunities by restricting them to a limited number of centralized, concentrated locations only, the policy may well worsen the situation

Plate 8 Abandoned market infrastructure: Colombo

of the smaller, most vulnerable operators. Inevitably these operators are in the least competitive positions in terms of appropriating one of a limited number of desirable sites. Usually, however, they are, in absolute survival terms, the most dependent on informal trading. The generation of ways of accommodating the smallest, most vulnerable operators throughout the hierarchical range of spatial opportunities must be an important part of market-policy formation (Figure 1 and Plate 7).

There are three interrelated factors which are essential for a successful markets policy. The first is a long-term political commitment. Central to the whole issue of creating opportunities for small-scale economic activity is creating confidence. This cannot grow when attitudes to informal-sector activity are frequently changing. The second is a clear unambiguous policy. Lack of clarity leads to confusion and conflict and, in the long term, it undermines confidence. The third is a regular and predictable flow of funds. The amount of finance necessary to promote market activity need not, as will be shown in subsequent sections, be great. However, it is necessary to think through a cohesive plan that provides a definable path and pattern of development. For this to occur, it is necessary that a regular flow of finance exists. Markets must be treated as an essential form of urban infrastructure – as essential as roads, schools, or other urban elements – and confidence will not be engendered adequately until local-authority investment patterns reflect this.

Location

The location of markets is a critical factor in their success. The analysis uncovered many cases where inappropriate location had led to the total failure of markets: in many cases, expensive shelter and infrastructure had simply been abandoned while vendors took matters into their own hands to find more suitable locations (Plate 8). At a city scale there are three major factors affecting the location of markets.

Location of generators of population movement

Markets are extremely sensitive to flows and concentrations of pedestrians and traffic and the most successful locations are therefore in close proximity to larger generators of population movement (Figure 2 and Plates 9, 10, 11, and 12). Markets therefore operate most successfully in central business districts and other formal commercial agglomerations, industrial concentrations, around public transport terminals (for example, bus terminals, train stations, taxi ranks, and particularly where two or more of these terminals are proximate), central locations in high-density areas, and so on.

By definition, the different generative capacities of different parts of the city give rise to a hierarchical range of locations with commercial possibilities. If unrestricted, agglomerations of small traders will collect at points where

Figure 2 The relationship between generators of population movements and small traders: (a) Colombo; (b) Pettah area of Colombo

At a city scale; selling from both formal and less formal markets is closely related to railway stations and bus termini.

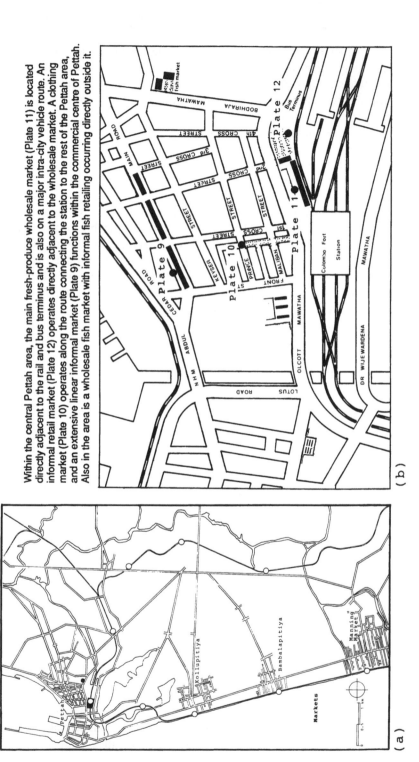

Within the central Pettah area, the main fresh-produce wholesale market (Plate 11) is located directly adjacent to the rail and bus terminus and is also on a major intra-city vehicle route. An informal retail market (Plate 12) operates directly adjacent to the wholesale market. A clothing market (Plate 10) operates along the route connecting the station to the rest of the Pettah area, and an extensive linear informal market (Plate 9) functions within the commercial centre of Pettah. Also in the area is a wholesale fish market with informal fish retailing occurring directly outside it.

Plate 9 Pettah Street market: Colombo
Plate 10 Informal clothing market near station: Colombo
Plate 11 Central wholesale market: Colombo
Plate 12 Informal retail fresh-produce market: Colombo

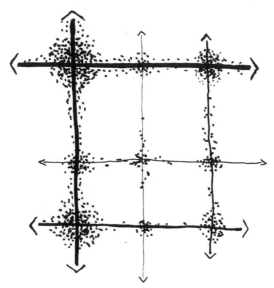

Figure 3 Larger agglomerations of traders collect at points where population movement is greatest.

population movement is greatest (Figure 3). For planning purposes, an understanding of the generative capacity of different parts of the city will give a guide to potentially good locations for markets as well as to their possible size.

Sources of supply

A second factor determining the location of markets at the city scale is the siting of major sources of supply – for example, easy access to the main fresh produce wholesale market is important for fresh produce retail markets. Proximity to wholesale supplies also has an important effect on whether or not the smaller, poorer trader can survive in a market. Thus, the further the retail market from its source of supply, the greater the tendency for the market to be dominated (frequently in a quasi-monopolistic fashion) by larger, more wealthy traders who are better able to bear the costs of overcoming the friction of distance. It is only the wealthier traders who can afford private vehicular transport to move their supplies, who can afford to bulk-buy and store without spoilage, and so on. Therefore, if one aim of a markets policy is to promote employment and improved incomes for the poorest section of a city's population, markets should, as far as possible, be located close to wholesale suppliers (for comment on the location of wholesale markets, see the subsection on supply-side support, below).

Location of consumers

From a planning point of view, a third factor which should influence decisions on market location is the need to serve the city's consumers as equitably as possible. While wealthier people generally have access to private transport and can easily satisfy their shopping needs, poorer people (because of their lower income and greater time constraints) are usually confined in their consumer behaviour to their local areas. For poorer people, access (by foot) to the range of basic and often cheaper products provided by a market can significantly improve the quality of their lives. Lower-income areas should therefore be well provided with markets which occur in a decentralized rather than a centralized form. The obvious corollary of this is that residential densities must be high enough to provide the thresholds necessary to support such a decentralized markets system: low densities actively promote centralization of economic activity, to the disadvantage of both operators and the people they serve. Conversely, one of the most important factors affecting the intensity of informal-sector activity is the compactness of the consumer base it serves.

At a more local level, markets are extremely sensitive to pedestrian flows. This sensitivity is, however, significantly affected by the size of the market. Where a market is made up of relatively few traders, a location even a few yards away from their optimum position can result in the failure of their business.

Plate 13 Informal selling for a formal market-building: Hong Kong

In this market in Hong Kong, sellers have turned their backs on the formal market in response to larger volumes of pedestrian flow along the street

Where agglomerations of traders are larger, however, they can act as an attractive force in their own right: pedestrians will move to them. As a rule of thumb, markets need to be fairly large (containing in excess of several hundred operators) before they have really significant independent drawing power, and can challenge or break the pre-existing pattern of pedestrian or traffic flow (Plate 13).

This realization emphasizes the need to think *processionally* about market provision in terms of permanence and scale. It is frequently impossible to predict, via analytical or quasi-scientific techniques, the probability of growth of a market: success breeds success and it is therefore necessary to think sequentially about the possibility of growth. Phased growth of a market must be designed in such a way that the market operates as a totality in each phase of its development and phased capital investment in infrastructure should occur (Figure 4). As a general rule, more expensive infrastructural investment should follow market development only as and when required: it should not precede it. Some of the most spectacular failures have occurred when the scale of the market has been overestimated from the outset.

At a local scale, the potential of different locations is not necessarily the same over different periods of the day. Market location, therefore, should not necessarily be seen in terms of a single location: it can also take the form of a system of locations, which allow operators to respond to variations in patterns of

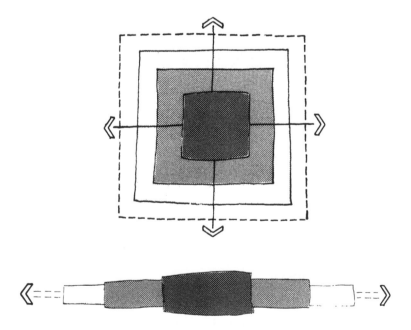

Figure 4 Markets should be able to expand and contract and still remain as a cohesive whole.

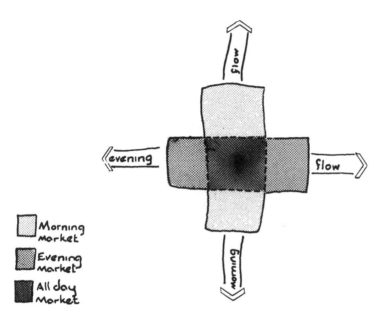

Figure 5 Market size and location can change over the day, depending on the strength and direction of pedestrian flows.

pedestrian flow over the day (Figure 5). This is the case particularly when a small number of vendors, engaged in trading products which are easily transportable, is involved.

Analytically, the best indicators of potential locations are provided by observing the locational behaviour of existing (frequently illegal) operators in an area: these traders are often the overt manifestation of a much larger, repressed, demand for space by small traders.

At a more detailed level, it is important that the market infrastructure is integrated into the surrounding urban fabric. Markets need not be 'disruptive' elements either visually or in terms of traffic and pedestrian flow, as is frequently feared by city officials and politicians. Nor should it be assumed that markets will 'downgrade' an area and cause it to appear 'untidy' or messy (Plate 14). Careful attention to market design in relation to its surrounds can resolve these difficulties. Thus, for example, even in built-up urban zones such as central business districts, cheap market structures with low levels of utility services can be inserted into the urban fabric in a non-disruptive manner, simply by designing the façade of the market entrance in such a way that the continuity and style of the street are maintained.

Plate 14 Markets need not be aesthetically negative or disruptive in terms of pedestrian flows: Bangkok

Use mix

The way in which different uses in a market (fruit and vegetables, meat, fish, clothing, and so on) are located relative to each other, and the way in which they are scattered or concentrated, are important considerations in the design and administration of any market system.

On a broader scale, the needs of consumers are best met when a wide range of products is available for sale in close proximity. In low-income areas especially, markets should be accessible to people who have to move primarily on foot. This does not necessarily mean, however, that each market should contain a wide mix of products.

The product mix of a market tends to be determined primarily by three interrelated factors. The first is city location (for example, markets in predominantly industrial areas frequently contain a high incidence of cooked food vendors, while those in central business districts more commonly reflect an emphasis on durable goods, such as clothes, and household goods). The second is the particular needs of the community which the market serves (thus, for example, in low-income communities, the most common products found are fruit and vegetables, poultry and meat, clothes and household goods, while markets in higher-income communities frequently emphasize more luxury items such as

Plate 15 Where appropriate, markets should create opportunities for small-scale manufacturing on site: Musika market, Harare

ornaments, pottery, and artwork). The third is the availability of alternative markets (the more the number of markets in any one area is restricted, the greater the range of products in any one market is likely to be).

Significantly, markets should not necessarily reflect an exclusively retailing function. Informal small-scale manufacturers (Plate 15) are frequently unable (because of the need to keep overheads to a minimum) to separate the manufacturing and selling functions, and thus thrive in a market type of location where both can be carried out simultaneously.

Generally, in terms of any one market, a multifunctional character increases drawing power, but products represented must attain a 'critical mass' – there must be enough of them to make a prominent presence and to have a collective drawing power – for them to contribute significantly to the activity of the market or for the vendors supplying them to do well. The greater the number of traders who are spatially separated from major pedestrian flows and therefore rely on conscious effort on the part of the consumer to move to them, the more the issue of critical mass is important. Perhaps the most glaring example of this, commonly found in formal market buildings in the Far East, is when products are vertically separated: unless there is a very strong concentration of vendors on the upper floor, few consumers can be enticed to ascend to it (Figure 6 and Plates 16 and 17).

An important dimension of functional mix is a recreational component (for example, refreshment vending associated with places for sitting and relaxed

31

Plate 16 Ground floor of Bambalapitiya market, Colombo

Plate 17 Deserted upper floor of Bambalapitiya market, Colombo

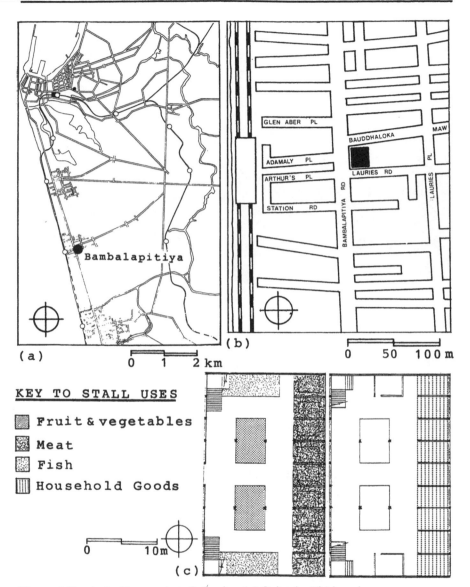

KEY TO STALL USES

▓ Fruit & vegetables
▓ Meat
▓ Fish
▓ Household Goods

Figure 6 Bambalapitiya market, Colombo: (a) Colombo; (b) Local area map; (c) Bambalapitiya market

This is an example of a market which fails because it is too small. The space provided for each category of use is so limited that none can attain the necessary 'critical mass' needed to attract significant numbers of customers. Thus, the ground floor of the market is 360 square metres in size and contains only eight fruit-and-vegetables stalls, four fish stalls, and nine beef stalls.

The upper floor of the market contains lock-up shops for household goods and clothing. Most of these, however, are closed: there is simply too little activity to induce people to walk up the stairs.

Plate 18 Refreshment stalls outside: Singapore

Markets often contain a recreational component.

gathering, play facilities, and so on [Plate 18]). In low-income communities particularly, the success of markets is strongly related to the degree to which they become a community social focus.

Perhaps the most important aspect of functional mix is the strong need for internal specialization: similar goods – for example, fruit and vegetables, fish, meat and poultry, clothing and ornamentation, household goods, and so on – need to be organized into functionally identifiable and discrete zones within the market. The need for this is clearly reflected in markets which have developed organically in relatively unconstrained circumstances: in almost every case a high degree of internal specialization has emerged.

There are five main reasons why this differentiation is important. First, markets made up of small vendors are dependent upon a high degree of comparative buying. This is the case particularly when the markets serve relatively low-income communities: because the opportunity cost associated with each unit of currency spent is high, great care is exercised by consumers in relation to even very small purchases. The process of comparative buying is greatly assisted by internal specialization, which clearly signals *where* consumers should be searching. In cases observed where individual vendors were dislocated from specialized areas and were spatially associated with different goods, the

34

Plate 19 Fish selling needs washing and drainage facilities: Bombay

Plate 20 Other products require a different selling environment from meat and fish: a pavement market, Colombo

Plate 21 This stall infrastructure is significantly non-specialized so as to allow the display of many different types of goods: Bombay

vendors affected were severely disadvantaged. Second, consumer behaviour is highly probabilistic: the likelihood of a customer using a particular vending outlet is related to the expectation of satisfying the purpose of the shopping trip. The concentration of particular goods and services affects the *image* of the market amongst consumers. Third, different types of goods have different requirements (such as loading) and sometimes require specialist services (for example, washing and drainage facilities are essential in the case of fish vending [Plate 19]): concentration facilitates the provision of these. Fourth, different goods have different externality effects (for example, smell and visual impacts) and these are not always compatible with other goods and services: there is a constant danger of cross-contamination if no spatial differentiation occurs. Fifth, different goods have different environmental requirements to optimize selling – for example, the light and other display requirements for clothing are totally different from those for selling meat (Plate 20). Internal specialization enables the creation of different retailing environments which suit the needs of particular products.

The need for specialization has implications for both design and administration of markets. From a design perspective, the first need is to minimize negative externalities, through the physical definition of the relationships between compatible and potentially incompatible uses, through the creation of clear defining edges between uses, and through careful physical resolution of the contact zone between incompatible uses. The second is consciously to encourage environmental conditions appropriate to the use in question. The third is that while specialized facilities and services are necessary for some uses, where possible, particularly in cases involving the spatial proximity of compatible uses, facilities should be as generalized as possible to enable expansion and contraction of the specialized area (Plate 21).

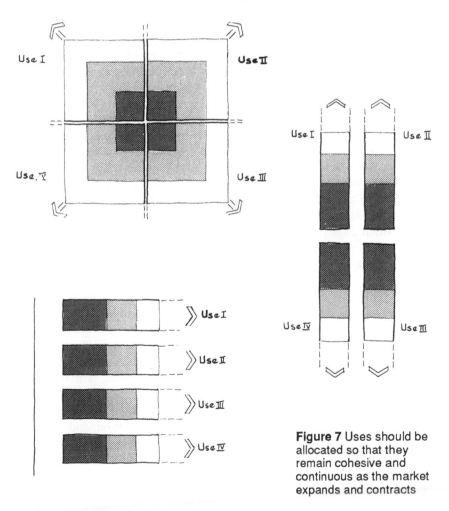

Figure 7 Uses should be allocated so that they remain cohesive and continuous as the market expands and contracts

37

From an administrative perspective, the allocation of sites within a market – particularly those that are pronouncedly periodic, in that they undergo considerable expansion or contraction over the day or week – clearly cannot be handled simply on a 'first-come, first-served' basis. A preliminary identification of zones of use is necessary and allocation needs to occur in such a way that the market remains as cohesive and continuous as possible in the process of expansion (Figure 7).

Physical layout

Markets may take many physical forms. Market performance, however, is affected by the following related issues.

Spatial marginalization

Probably the most common problem which is directly related to physical layout is that of 'spatial marginalization'. This may take the form of areas in a market in which stalls are unused or abandoned, areas in which stallholders are generating low profits relative to other stallholders in the market, and areas which tend to be avoided by customers. This phenomenon is commonly found in planned, formal market buildings. Traders are not adequately exposed to sufficient flows of potential custom, reflected in pedestrian movement, for them to operate viably. The best situation obtains when intense flows are diffused across the entire trading area, and where 'edge' or cul-de-sac conditions are reduced (Plates 22 and 23). Conversely, the greater the extent of 'dead spots' within the market, the poorer and more inequitable its performance. Significantly, too, the greater the extent of differential performance within a market, the greater the administrative problems which ensue: competition for more viable locations is intense and frequently takes socially destructive forms (violence, bribery, and so on). Further, tendencies towards monopolization by stronger, more viable operators are aggravated.

The diffusion of pedestrian flows across a market area is primarily affected by three design factors. The first is environmental. Capitalization on the different selling environmental requirements of different uses, and the strategic relational alignment of those uses, can be used to 'draw' customers through the entire market area. A good example of this is to be found in Manning market in Colombo (Figure 8). Immediately inside the market entrance is a fruit and vegetable section under a double pitch roof. This section of the market opens on to an open-air courtyard around which are meat and fish stalls. The fact that the open-air courtyard is partially visible from the enclosed section, and presents itself as a very different kind of environment, draws customers through to this part of the market and prevents stallholders around the courtyard from being spatially marginalized. A related issue is making maximum use of generative activities with the market. Certain types of goods or scales of operation have

Plate 22 Formal market-building: Hong Kong

Plate 23 Open-air market: Colombo

These photographs show 'dead spot' situations. Plate 22 shows a cul-de-sac problem: there is no through-flow of customers and shops are mostly closed. Further, there is excessive circulation space and 'dead edge' backs of stalls kills the intensity of commercial activity. Plate 23 shows a cul-de-sac condition in an informal market set up by the municipality.

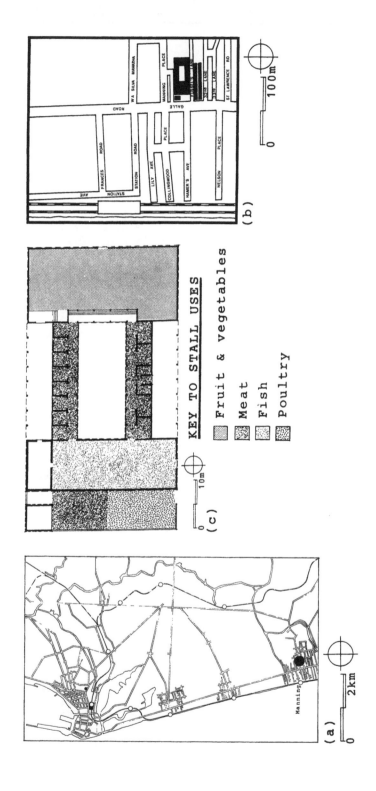

KEY TO STALL USES

Fruit & vegetables
Meat
Fish
Poultry

(a)

(b)

(c)

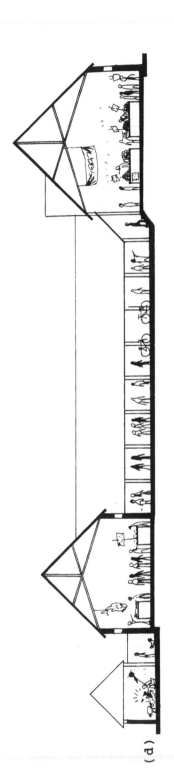

(d)

Figure 8 Manning market, Colombo: (a) Colombo; (b) Local area map; (c) Manning market; (d) Manning market cross-section

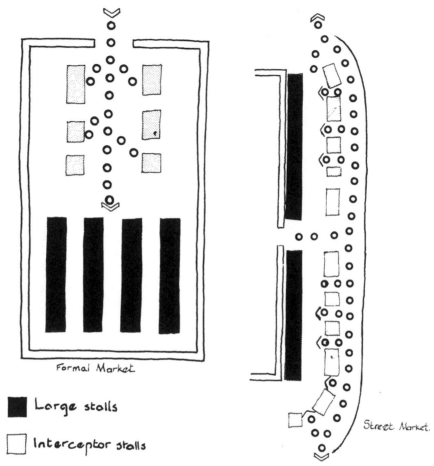

Formal Market

■ Large stalls

□ Interceptor stalls

Street Market.

Figure 9 Smaller stalls survive by intercepting the flow of customers moving through to larger stalls

greater drawing power than others. Some of the most successful markets occur when smaller operations or those with less generative power are orientated directly to the major movement-flows and the stronger ones are located behind them. The stronger elements draw custom through the market while smaller ones adopt an interceptor position in relation to that custom (Figure 9).

The second factor affecting pedestrian flows is the orientation of the market to dominant pedestrian circulation patterns (Figure 10, a–d). In behavioural terms, consumers seek to reduce energy expenditure while searching for purchases. They almost always gravitate towards the most direct movement channels (the passageways between stalls or shops) in their pattern of search. When the orientation of market stalls to, and the integration with, these flows is close, spatial marginalization is reduced. Conversely, the greater the lateral spread of

Plate 24 Formal market building: Singapore

Stalls are orientated at right angles to the main circulation routes, thus exposing primarily their outer edges to the main pedestrian flows. Inner stalls are more frequently unused.

selling areas away from the dominant movement channels, the greater the tendency towards marginalization. In the case of less formal markets, these flows tend to be defined by urban circulation channels (particularly streets or pavements) and for this reason, the form of informal markets is almost invariably linear: sellers seek to locate as close as possible to these flows of potential customers. In the case of nucleated markets (and most markets accommodated in formal market structures or buildings take this form), the directional emphasis of pedestrian flows is defined by the pattern of entrances to, and exits from, the market. In less formal nucleated cases, access is usually more porous than in more formal ones: people can move into the market at a number of points. In the latter case, the main direction and volume of pedestrian flow into the market is determined by the relative importance of the urban movement channels which flank the space. In the former the *precise* pattern of entrances is the major determinant. The resolution of the relationship between entrances and exits, circulation space, and selling area thus becomes critical. The worst cases of lateral marginalization occur when the longer selling runs are orientated not with, but across, major movement flows (Plate 24).

The third is visual contact. The propensity for a part of a market to be used is strongly related to the degree to which it is visually observable from other parts

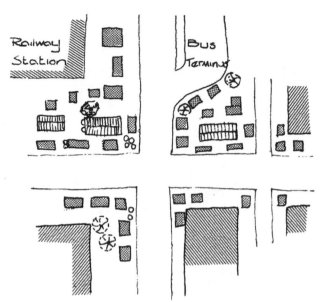

Figure 10(a) Informal traders will always attempt to establish themselves at points of highest pedestrian concentration. Here they form linear markets aligned to pedestrian flow.

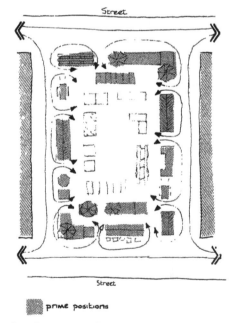

Figure 10(b) Informal traders occupying a square or rectangular site will gravitate towards the edge of the site so as to be closer to passing pedestrians.

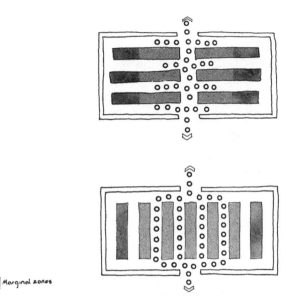

Marginal zones

Figure 10(c) In a formal market-situation, pedestrian flows are affected by stall orientation and entrance location. In these two examples, the pedestrian flow is unevenly distributed. Peripheral stalls are marginalized and become 'dead areas'. (See Plate 24).

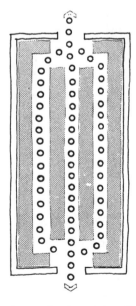

Figure 10(d) The stall orientation and entrance location distributes pedestrian flows evenly past all stalls and there is no marginalization or 'dead areas'.

Plate 25 Informal market: Colombo

In this case, stalls are allocated on a semi-permanent basis. The market has, however, contracted during a non-peak period, leaving certain sellers isolated from customer flows.

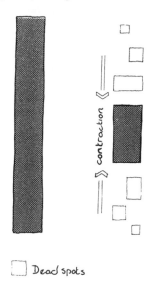

Dead spots

Figure 11 Informal market: Colombo

of the market. This is particularly important in markets which occur in formal structures and in which there may be vertical separation of different uses.

The issue of marginalization is also affected by the scale of the market. The larger the number of purchasers using the market, the greater the tendency for 'lateral spill-over' towards less direct movement channels to occur. Four other

Figure 12 Linear selling runs intersecting around a number of 'cores'

forms of 'dead spots' which need to be consciously overcome are commonly observable in markets.

The first is 'dead spots' caused by a non-contiguous, fragmented market form. Because there is frequently a high incidence of periodicity, particularly in informal markets, there is a danger that, in the daily process of market contraction at non-peak selling periods, certain vendors will be stranded, in the sense that they are removed from the mainstream of market activity (Figure 11 and Plate 25). Severe economic disadvantage accompanies this phenomenon. This, together with the need to maintain a degree of product specialization (discussed above), has considerable implications for physical layout. Some of the best markets can be conceptually conceived of as a number of product-specialized, linear selling runs (i.e. lines of adjacent stalls) which intersect in complex ways in or around one or a series of cores (Figure 12). This allows contraction and expansion to occur over the day, while maintaining definable areas of product emphasis and a cohesive market experience at each stage of development.

The second form of 'dead spot' occurs when formal shops or kiosks are located on the edges of nucleated markets housing informal operators (Figure 13). This form of layout almost never works. The trading focus is inevitably the informal market and this establishes the tempo of activity and the dominant trading environment. Small shops, almost by definition operating more individually than the collective market in terms of drawing custom, simply cannot compete.

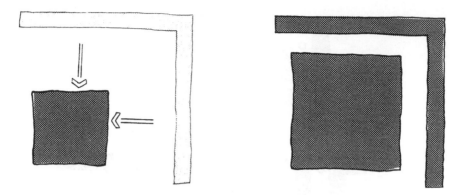

Figure 13 'Dead spots' caused by market contraction away from peripheral formal shops

The third form is 'dead spots' which occur around the middle of excessively long, unbroken, selling runs (Figure 14). Purchasing behaviour within markets is primarily influenced by two potentially conflicting consumer motives: comparative buying and convenience. Because of this, there is a definable pattern of market intensity which peaks at the intersections of selling runs and cross-circulation channels and declines with distance away from these points. When the length of selling run between cross-circulation channels is excessively long, dead spots tend to occur towards the middle of the run and the incidence of vacancies is usually highest here. Observation shows that this occurs when the run is in excess of 50 or 60 metres.

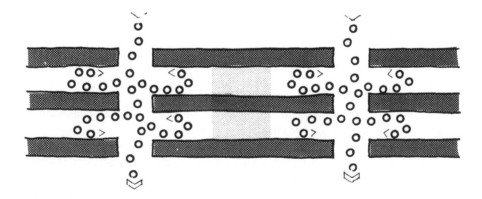

Figure 14 'Dead spots' at the centre of excessively long runs of stalls

Plate 26 'Dead spots' caused by the non-selling sides of stalls: Hong Kong

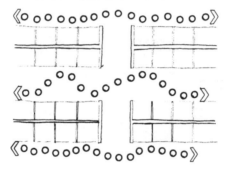

Figure 15 'Dead spots' caused by the non-selling sides of stalls: Hong Kong

The fourth form of 'dead spot' is caused by the non-selling sides of stalls or spaces within a market (Figure 15 and Plate 26). Vendors occupying corner sites of intersecting circulation channels frequently sell from one edge of a stall or space only, particularly when the total area available to them is small (for reasons of inadequate space) or when they are operating individually (because of inadequate manpower). Logically, in these cases, they will orientate the selling surface to the major movement flow, thus presenting a dead edge to the cross-circulation channel. When the spacing between channels is too short, the combined dead effect of the two corner sites creates a feeling of commercial deadness, thereby repelling custom and negatively affecting the economic performance of interstitial vendors.

Length of selling runs

Market performance is significantly affected by the length of unbroken runs of adjacent stalls. Two issues are central to this. The first is that unbroken runs (that is, the selling areas between two cross-circulation channels) must be long enough to facilitate comparative buying and to generate a strong sense of vibrancy and activity. When the runs are too short, activity levels are dissipated: the amount of

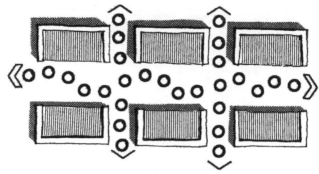

Figure 16 Selling runs are too short and customer flows are dissipated and confused

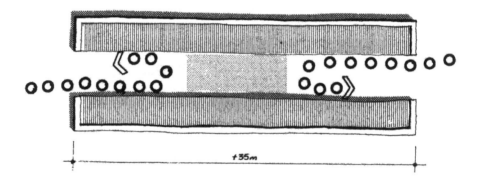

Figure 17 Selling runs are too long and customers do not penetrate to centrally located stalls

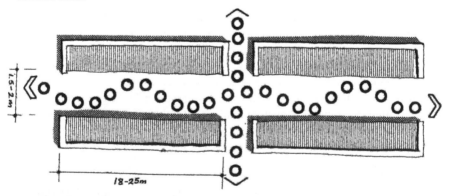

Figure 18 A more appropriate length for selling runs

circulation space relative to selling space is increased and this has a deadening effect on activity (Figure 16). Further, searching and comparison is difficult and

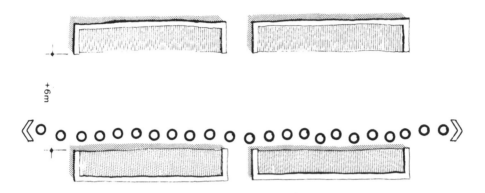

Figure 19 When circulation channels are too wide, customers concentrate on one edge only

Plate 27 Informal market: Colombo

In this market, circulation space is approximately 14 metres in width. The two sides of the market operate independently and customer flows are diffused.

market space is wasted, since the ratio of circulation to selling space is unnecessarily high. Conversely, when the runs are too lengthy, the ability of consumers to switch between runs is impaired and selection is again restricted (Figure 17). As discussed above, too, when the run is excessively lengthy, commercial dead spots tend to occur towards the centre of the run.

Observation shows that the best situation seems to occur when the runs are in the order of 18 to 25 metres: this rhythm is frequently encountered in spontaneous markets which have developed organically in situations where there has been

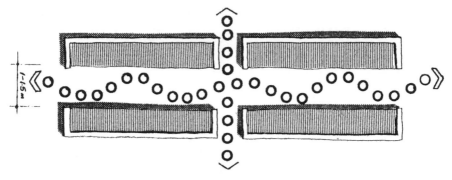

Figure 20 A more appropriate circulation-channel width

little or no constraint on their formation (Figure 18). Problems occur when runs exceed 35 metres or are less than 10 metres in length.

Width of circulation space

The width of circulation channels (i.e. passages between rows of market stalls) affects how markets operate. The optimal situation pertains when purchasers are able to engage vendors on both sides of the circulation channel in the process of product selection. When the circulation space is too wide, consumers concentrate on one edge of the channel only: the two sides operate independently (Figure 19 and Plate 27). Frequently, in these situations, vendors on one side of the channel will markedly out-perform those on the other. When the channel is too narrow, the resultant congestion impedes the shopping function.

Clearly, there is no one optimal width of circulation space: it varies with market intensity, design, use structure, and so on. A clue to the range of dimensions involved, however, can be gained through observation of vendors in spontaneous markets. In intense pedestrian conditions, vendors consistently attempt to define channels of 1 to 1·5 metres in width (Figure 20). Where passages are narrower than this, congestion becomes excessive and there are usually observable attempts to widen the passage. Once the circulation space exceeds 3·5 to 4 metres, however, there is a tendency for vendors to seek to divide pedestrian flows by taking up locations in the centre of the channel (Figure 21 and Plate 28). In design terms, a central issue which emerges from this is how to create the possibility of expanding the volumes of spaces with changing intensities of pedestrian flow over different times of the day or week.

An important issue relating to movement circulation, too, is how to reduce conflict and congestion caused by directional shifts of pedestrian movement. This is most successfully resolved by creating expanded 'knuckles' of space or absorption spaces at the intersection of movement channels. Where these exist they operate not unlike urban squares or public spaces in a city, but on a very

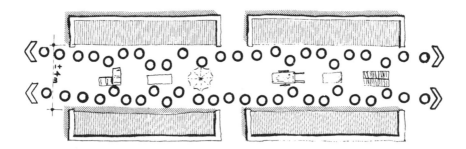

Figure 21 Both Plate 28 and Figure 21 show a situation where sellers have spontaneously narrowed a wide circulation passage by locating in the centre and dividing it in two.

Plate 28 Street market: Hong Kong

small scale: they provide places where people can pause, or shift direction, or stand and talk in the case of chance meetings between acquaintances.

Market infrastructure

Internationally, a wide range of types and levels of market infrastructure is to be found in urban markets, ranging from situations where nothing is provided by the authorities (for example the open markets found in many parts of Asia and Africa, in which product display occurs from the ground) to situations where high levels of infrastructure are publicly provided (for example, the market buildings,

Plate 29 This night market occupies an entire street: Singapore.

fully serviced and built to high standards of finish, of Singapore, Hong Kong, or Taipei). There is no one 'best' level or form of market infrastructure, however. The following general observations can be made about the issue of infrastructural provision.

First, the economic success of markets rests primarily on their location and their environmental quality. That quality is defined by 'vibrancy', colour, hygiene, and the spatial experiences of users. It is not necessarily related to the level or type of infrastructure. Indeed, some of the least successful markets found in Asia occur in extremely expensive, highly serviced, formal market buildings.

Second, the higher the level of infrastructure, the higher the cost to the users, if the market authority is working on a cost-recovery or a profit basis. Given this, as levels of infrastructure increase, there are increasing tendencies to price out smaller, more vulnerable, traders: a desire to increase infrastructural levels may, at a point, conflict with the *social* purpose of a positive markets policy (which is creating trading opportunities which are equitably accessible to all). In fact, almost nowhere do fully serviced market buildings, serving low-income communities, pay for themselves. Moreover, in most cities the subsidy level is increasing: the level of subsidy is relatively much higher on buildings built today than on those constructed in the past. Cities which espouse policies of full infrastructural provision do so for social or political reasons: nowhere have they managed to recover the costs totally and the cost of subsidy is perceived as a

Plate 30 A pavement market: Bangkok

necessary social payment. Significantly, too, policies of full infrastructural provision are being reconsidered in almost all cities which previously held them.

Third, the type and level of infrastructure which is appropriate is informed by the permanence of the market: the greater the degree of permanence, the greater the extent of fixed infrastructure which can be considered. Conversely, in cases where markets are highly periodic, it does not pay to provide any fixed infrastructure.

Fourth, markets are frequently periodic: they often operate at certain times of the day, week, month, or year only. In these situations, the multifunctional use of urban spaces and elements is essential. Thus, some of the most successful markets are found in urban streets (Plate 29), pavements (Plate 30), squares, and parks (Plate 31). Frequently, other activities (for example, vehicular movement in the case of streets or squares, or leisure activities in the case of squares and parks) can occur simultaneously, with only the emphasis of use changing at different times of the day. In these cases, the different activities complement each other, either by increasing vibrancy or by providing an easy temporary escape from the intensity of market activity. An interesting example of the multi-functional use of urban elements can be found in Colombo, where vending kiosks are attached to a school wall: the school, which is anyway a focus of community life and activity, contributes to the generation of market custom, while market activity and exposure contributes to the informal education of children.

Plate 31 Market in a park: Bombay

Plates 32 - 4 Multi-use of street space: Singapore

A street in Singapore which functions as a through-route during the day. At night, it is closed off and turned into a street market. Plate 32 shows how stalls are erected from light bamboo and canvas in the evening. Electricity connections are made to adjacent shops and residences, and the market attracts large volumes of customers until near midnight. Thereafter, the stall-holders clear up and dismantle their stalls, and the road is open for traffic again the next morning.

57

The common dimension of periodicity characterizing various urban functions also enables different urban spaces and elements to be used differently over different time periods (Plates 32, 33 and 34). Thus, there are many examples where roads are temporarily closed to become street markets; in Singapore, central-city parking areas become food centres at night; in Bangkok, storage sheds are used as market structures at certain times; in Lima (Peru) the central-city pavements are turned over to vendors at night; and so on.

Importantly, therefore, markets should not be viewed as environmental nuisances which should be tolerated or suffered. If properly managed, they offer considerable environmental potential and can be consciously used to improve environmentally sterile urban 'dead areas' (for example, they can be used to give scale and definition to excessively wide street or pavement spaces, to give life to areas like parking spaces, and so on).

Finally, people frequently show great ingenuity in providing much of their own market infrastructure (particularly shelter and selling surfaces). Some of the best environmental situations result when this creativity is released. The propensity for people to make this contribution, however, is related to two factors. The first is the selling period. When vendors are selling for very short periods of the day (for example, one or two hours) they will set up infrastructure which demands as little effort as possible (Plate 35). The second is security. When vendors are operating in insecure conditions and are subject to official

Plate 35 Movement of this stall infrastructure requires very little effort: Bangkok

harassment, they invariably provide very little infrastructure. The creation of a sense of security, therefore, is an important dimension in encouraging people to contribute to their own business environment. More detailed observations on individual elements of market infrastructure follow.

Cleanable floor surface

The existence of a hard, easily cleanable market floor surface is a considerable advantage, particularly in wet climates. Two sets of problems result if the surface is not appropriate. The first are problems of dust and mud, which make display difficult. The second, found particularly in fresh-produce markets, is smell. There is a tendency for produce to be trampled into the ground and when easily cleanable floor surfaces do not exist, the material leaches into the soil and decays, creating unpleasant odours which do nothing to enhance the trading environment.

Water

The presence of potable water at or near the market is important and serves three functions: it enables regular cleaning, both of the market area at large and of individual selling areas; in the case of food markets, it allows produce to be

Plate 36 Communal taps are sufficient in some produce markets: Musika market, Harare

washed regularly, which is important both for hygiene and for attractive display; and it provides drinking water for vendors and customers. Frequently, in markets at which water is not available, vendors provide their own and store it in drums or tanks. Often, the water is not regularly replaced and the standing water can become a health hazard.

The amount of water provision necessary varies with the type of goods being sold. In the case of fish, meat, or freshly slaughtered poultry, easy access to water is essential: ideally, there should be at least one tap per two or three selling spaces. In the case of most other products, communal taps are sufficient (Plate 36).

The cost of water provision can be recovered through metering where supply is to the individual trader or a small group of traders. In the case of communal taps, it is impossible to apportion user responsibility and in practice the cost must be carried as a public overhead. Provision of coin-operated meters simply discourages use.

Plate 37 Public provision of electricity – metered connections to stalls: Bangkok

Electricity

The need for access to electricity varies with the type of market. In certain kinds of markets, particularly night markets and markets with a significant small manufacturing sector, it is essential. Similarly, certain types of traders, such as cooked food vendors, benefit from it: need can only be determined once the functional mix of the market is clear.

There are many ways in which electricity can be provided. Two forms are most commonly found in the Far East. The first is the public provision of metered standards at strategic locations, from which leads can be run (Plate 37). Frequently, a system of coin-operated connection exists and the cost of extension from the standard to the selling area is borne by the vendor. The second, which is frequently encountered in street or pavement markets which are physically proximate to houses or shops, is the private sale of electricity to vendors. Residents of the houses or shops allow vendors to run connections into their electricity supply for a negotiated fee: both parties benefit. Provided accountability is clearly established and the form of connection is not socially irresponsible, in the sense that it represents a real and obvious threat of personal hurt through electrocution or fire, there is no reason for authorities to prevent this practice. Indeed, it should be encouraged.

Plate 38 Simple shelter made from wooden poles and canvas: Harare

61

Plate 39 Shelter in the form of canvas umbrellas: Bangkok

Public toilet facilities

In situations where there is a high intensity of market activity and where alternative public facilities are not easily accessible, these are essential. The appropriate form and technology of provision will vary with context. The capital cost of such facilities cannot reasonably be carried by market vendors since the major users are the public at large. Further, direct user charges are, in most cases, counter-productive, since the main purpose of providing the facility is the promotion of hygiene. User charges simply discourage use of the facility and public or private open spaces are used. Capital costs should therefore be covered by general city taxes. Maintenance and cleaning costs, however, could reasonably be included in the rental charge imposed upon vendors. If this occurs, the market sellers will, themselves, find ways of keeping these costs as low as possible.

Plate 40 Shelter in the form of iron kiosks: Hong Kong

Shelter

A wide range of forms of cover and shelter can be considered. Common forms found in cities in many parts of the world include arcades; urban streets or walkways covered by the extension of some light-permeable material between buildings; suspended canvas; trees and other forms of vegetation; umbrellas; open-sided market structures; sheds; and formal market buildings (Plates 38–42).

Five main factors bear on decisions about the level and type of shelter which is appropriate. The first is *climate*. The primary purpose of providing shelter is not simply to increase the comfort of vendors, although this is a factor. It is to create an environment which enriches the trading experience and thus enhances trade itself and which allows the activity to break a dependence upon the weather. Local conditions of rain, temperature, and wind, therefore, affect the definition of appropriateness of shelter provision.

63

Plate 41 Shelter in the form of roofing over an arcade: Bombay

The second is *urban context*. The form of shelter provided should be appropriate to, and should enhance, the local context of the market and should, where possible and for reasons of economy, make maximum use of locally available opportunities (for example, the use of trees for suspending canvas in park situations; covering narrow walkways; and so on).

The third is *environmental impact*. The environment of a market is a major factor affecting its economic success. Environmental quality, in turn, is primarily affected by spatial qualities (which are themselves informed by market form) and by the quality of light: excessive gloom (or glare in very hot climates) has a major dampening influence on market activity. The type and form of shelter provided significantly affects this, and even though vendors frequently exercise great ingenuity in creating their own shelter, on occasion design research to enable the initiating authority to advise on this is warranted.

The fourth is *market permanence*. The more permanent the market, the greater the range of shelter options which can be considered. Conversely, the greater is

Plate 42 Shelter made from canvas-covered steel poles: Bangkok

the periodic dimension to market activity, the less feasible it is to consider permanent forms of shelter and the less willing traders are to provide their own shelter.

The final factor is *cost*. Obviously, the higher the level of shelter provided, the higher the cost: at the upper ranges of provision, the cost of shelter is the largest item of market capital-expenditure.

Ultimately, there is no standard answer to the question of whether the public authority or individual vendors should be responsible for the provision of shelter: it requires contextual assessment. People are generally able to exercise great ingenuity in satisfying their own shelter requirements. The role of the authority may be simply to initiate certain 'enabling actions' to allow this to occur, for example, the provision of posts or struts to support canvas; assisting access to materials; drilling holes to take large umbrellas; and so on. Direct intervention by authorities may be necessary when there are income constraints (when a significant proportion of traders cannot provide shelter for themselves); environmental considerations (when particular treatments suggest themselves as a means of benefiting the market as a whole); technical considerations (for example, when structural work is required); and urban design considerations (when a cohesive standard approach is necessary to maintain the environmental quality of an area).

Plate 43 Display on bare earth: Harare

Plate 44 Display surface of concrete slabs: Harare

Plate 45 Display surface made from boxes: Hong Kong

Selling and display areas

The ability to display goods and produce adequately and attractively is an essential part of small-scale vending: normally low-income people are highly selective in their buying habits and considerable competition occurs between traders in the arena of display. In the case of food markets, clean selling surfaces are also important for reasons of hygiene.

Many forms of display space, ranging from those which are entirely privately provided to those entirely publicly provided, can be observed internationally. They include: earth surfaces marked on the bare ground; surfaces made of boxes, packing cases and other transferable materials; wooden-framed stalls (all privately provided), publicly provided support struts or poles between which surfaces can be suspended (public–private provision); permanent plinths made of durable materials, extended slabs of concrete, or brick, formal kiosks (all publicly provided); and so on (Plates 43 – 5).

Different uses have different display and selling requirements. The sale of piecemeal goods requires flat surfaces upon which articles can be grouped and arranged. Fruit and vegetable traders require all products to be displayed. They need to be able to assemble variable quantities of goods easily and thus they often need space to store fairly large quantities, and they also need to maintain close control over their goods. For all these reasons, expansion tends to occur upwards:

67

Plate 46 Fruit-and-vegetables stall: Hong Kong

fairly large quantities can be stacked in small areas (Plate 46). Meat, fish, and poultry require easily cleanable surfaces and in the case of meat, places to hang the product. Selling surfaces should ideally have drainage channels which connect with the broader market drainage system (Plate 47). In the case of fish, clogging of downpipes with scales is a perpetual problem if drains are too narrow. Clothing requires maximum display, and hanging facilities are a great advantage. A special case is provided by cooked-food sellers, who need a preparation area, a selling surface, and, ideally, seating facilities and eating surfaces for customers.

In some of the remarkably successful cooked-food centres of Singapore, the issue of table and chair provision led to considerable conflict. Initially, these were privately provided by stall-holders. The furniture itself, however, became an instrument of competition. If demand on one stall at any point in time outstripped its seating capacity, neighbours refused to lend, even on a temporary basis, unused tables and chairs, in the hope of drawing off the custom. Similarly, conflicts developed between stall-holders and consumers who lingered. The response of the Singapore authorities was to provide fixed tables and chairs which were communally available. The response has resolved some of the problems but the rigidity of the furniture arrangement and particularly the inability to move furniture in response to differing social circumstances impairs the social operation of the centre and thus its atmosphere.

Plate 47 Fish stall: Hong Kong

In terms of most products, the minimum necessary selling area is very small – of the order of 1 by 1·5 square metres. Generally, the smaller the selling area, the greater the market intensity. A particular issue which can cause problems is that of stall expansion at particular times of the day. The two most common forms of expansion during peak periods are outward into circulation space, which may create an unacceptable level of congestion and lateral expansion into other stall spaces. In fact, there is frequently pressure by stronger stall-holders to expand in this way and therefore to take increasing control of the market. This tendency should be curbed. In order to minimize both of these problems caused by expansion, two sets of stall sizes (a minimum and a maximum) should be designated to provide some flexibility while still maintaining control; the market must be able to operate well at both scales.

Storage

The provision of lock-up storage facilities in markets is problematic. Although they are frequently almost automatically provided as part of basic market infrastructure, often the facilities are not used, particularly by smaller vendors. Three concerns seem to underpin this. The first is that vendors feel that the level of security provided (usually simple lock-up spaces) is insufficient. Second, there is usually a rental cost associated with the use of the facility, and traders can

69

frequently find cheaper alternatives (for example, storing at home, coming to an arrangement – at very little cost – with nearby shopkeepers, using cheap lock-up facilities in railway stations, and so on). Third, it is often felt that the form of facility promotes product deterioration through inadequate air circulation, damp, or cross-contamination of perishable produce. In most cases, therefore, storage facilities should only be provided when an expression of real need emerges from traders: its provision should follow demand, not lead it.

The need for storage is greatest in the case of somewhat larger traders selling durable goods. The most successful forms of provision occur when the provision of storage is integrated with the selling function: when the need for storage is great, an alternative market *form* may be indicated (for example, wall kiosks [Plate 22] or lock-up kiosks in Hong Kong [Plate 40]).

Cleaning and garbage removal

The issue of cleaning and garbage removal is one which looms large in the concerns of local authorities, and this, together with the related issue of hygiene, is frequently quoted as an argument against informal trading from public places.

Observation indicates that the concern is frequently overstated. It is basically in the interests of traders to promote a pleasant and hygienic trading environment. Indeed, cleanliness is frequently an aggressive arena of commercial competition in urban markets. If the basic management conditions are correct it need not be a problem.

Three principles underpin responsible management in this regard. The first is clear definition of responsibility: arenas of responsibility between the market agency and the traders must be clearly spelt out. Ideally, traders should be responsible for the gathering of garbage, the removal of garbage to pick up points, and cleaning of the market area. The local authority should be responsible for garbage removal. When the definition of roles is not clear, no one takes responsibility and problems arise. Thus, the most common area of garbage build-up in markets occurs in the interstitial or semi-public spaces: traders keep both their immediate trading areas and (if they know it is their responsibility) the main circulation spaces swept and clean, but a build-up of dirt occurs in those spaces for which no responsibility has been assigned. A particular case emphasizing the need for the clear definition of responsibility can be found in the cooked food centres in Singapore. After problems arose with the private provision of furniture by vendors, fixed tables and chairs were provided as part of market infrastructure: customers could sit anywhere and choose food from any one, or a number of different, vendors. Once people had finished eating, however, cutlery and crockery would be cleaned by the vendor who owned them but the table surface remained uncleaned: no one was prepared to take responsibility for this. Eventually, permanent cleaning staff had to be appointed and this substantially increased market overheads.

The second principle is accountability. Since failure to carry out responsibilities in relation to cleaning and garbage removal by any one vendor can affect the trading environments of other vendors, a mechanism of enforcement must exist. Probably the most successful form of this is to make future market entry conditional upon responsibilities being met.

The third principle is the promotion of organization amongst vendors. When vendors enter into a formal and democratic organization, communication and negotiation between them and authorities is easier. A collective will is applied to conditions affecting trading and this leads to better trading conditions and environments; important collective back-up services (such as vehicle pools or other ways of simplifying wholesale buying, if this is necessary) can be organized; the chances of the market being dominated by one or a limited number of stronger traders is reduced; and internal enforcement of responsibilities is assured. Issues such as cleaning and garbage-removal responsibilities should ideally be negotiated between authorities and elected representatives of the collective body.

The issue of rubbish is also affected by market design. It is extremely important to create easy access to pick-up points on the periphery of the market. When the removal of garbage to these points is too arduous or time-consuming, tendencies to evade responsibilities by traders are encouraged. For this reason, a build-up of rubbish is frequently found at closed ends of market runs, market culs-de-sac, and other forms of dead-ends.

Supply-side support

A vital factor affecting both the extent and the viability of small-scale vending is the relationship between vendors and their primary sources of supply. Of particular importance in this regard is the urban fresh-produce wholesaling system, since basic foodstuffs arguably constitute the primary products sold by small-scale vendors, particularly in low-income areas. The central lesson which emerges from an examination of urban food-market systems is the importance of a spatially decentralized wholesaling system.

Importance of a spatially decentralized wholesaling system

To explain its importance, this issue needs to be discussed from three perspectives. The first is the relationship between wholesaling and small-scale agricultural activity. Intensive farming on small land units within or on the periphery of large urban areas is a characteristic of many countries, particularly developing ones. Frequently, this type of activity plays a central economic role for low-income families, constituting either the primary or a vital supplementary source of income. The viability of this activity, however, is strongly affected by access to markets and thus by the wholesaling system. The smaller the scale of agricultural production, the more significant the unit cost of transporting produce

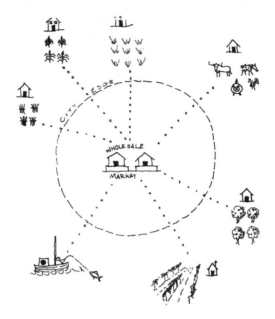

Figure 22(a) A centralized wholesale market increases the distance between producer and wholesaler

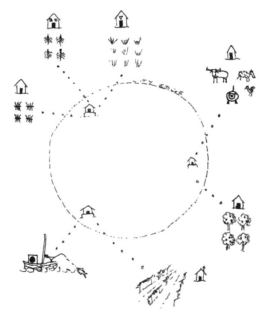

Figure 22(b) A decentralized wholesale system shortens distances between producer and wholesaler

to market becomes. By definition, the more spatially centralized the wholesaling system, the more small-scale producers are distanced from their primary markets. Conversely, the more direct the connection between producers and consumers, the better the requirements of small producers are met (Figure 22). The directness of connection is clearly strongly informed by city form and structure. The lower the density and the more sprawling the urban form, the greater the distance between producers and their main market. Ironically, too, the more sprawling the form, the greater the pressures towards spatial centralization of the wholesaling system.

The implication of this is that urban sprawl promotes monopsonic conditions of fresh produce supply and may completely destroy small-scale producers. As small producers become increasingly unable to afford the costs associated with frequently transporting small quantities of produce for considerable distances to centralized market points, middlemen arise to service the needs of a number of these producers. Because of their monopsonic position, they are able to engage in considerable price exploitation, to the detriment of both producers and consumers. Alternatively, small producers are forced out of business and the market gap is increasingly filled by large-scale producers who, however, require successively larger markets: subregional, regional, and even national systems of supply begin to emerge. These tendencies towards monopolization again work to the detriment of urban consumers and particularly the urban poor. Tendencies towards economic monopolization, in turn, encourage spatial centralization of the food wholesaling system: it suits a small number of producers controlling a large volume of supply to be able to deliver to one or a very limited number of distribution points.

The second perspective on the spatial location of fresh produce wholesaling is the relationship between wholesaling and small-scale vending. Easy access to wholesale produce is a vital factor affecting both the number of small-scale food distributors and their profit levels. When close connection exists, the cost of produce to the vendor is relatively lower and the cost of transporting supplies back to local market areas is reduced (Figure 23). Further, the ability to obtain additional supplies easily in the case of unanticipated demand increases buying flexibility. These factors in combination increase the ability of vendors to manoeuvre and thus, ultimately, to survive. For this reason, wholesale markets frequently have retail markets attached to them: vendors break bulk and sell primarily to other small-scale retailers, although final consumers also frequent them directly. Conversely, when easy access does not exist (and the greater the degree of centralization, the less the aggregate ease of access), flexibility and manoeuvring space are reduced, supply costs increase, profits decrease, and food costs to the consumer rise. An important structural change occurs in this process. Larger, more successful vendors can overcome more easily the costs associated with a lack of easy access: frequently, access to vehicular transport becomes a critical issue in the struggle to survive. This dramatically raises the capital entrance ceiling into vending and there is a definable pattern of economic

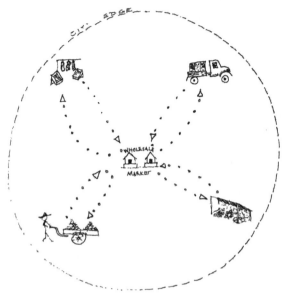

Figure 23(a) A centralized wholesale market increases the distance between wholesaler and small retailer

Figure 23(b) A decentralized wholesale system shortens distances between wholesaler and small retailer

mortality, with smaller, frequently periodic, traders collapsing first. The market is thus increasingly dominated by larger operators, tendencies towards monopolization accelerate, and both very small traders and consumers are the losers.

This point underlines the need to integrate the location of wholesaling outlets with public transportation. Large numbers (in many cases, the majority) of small vendors have no access to private vehicular transport. By definition, therefore, trading flexibility is greatly increased when the needs of hawkers and vendors are acknowledged within the public transportation system and when the system connects wholesaling and retailing markets. Thus, for example, in Bombay, India, the needs of vendors are considered, in part, by the practice of reserving the last carriage of every train for vendors' wares. All too frequently, however, the location of outlets is determined exclusively by the requirements of wholesale suppliers. In the case of retail markets which are spatially removed from wholesaling outlets, there is a strong case for the promotion of collective transport pools, in which vehicles can be shared among a number of traders, buying of supplies co-ordinated, and market entry prices thus reduced.

The third perspective on the location of wholesaling is the intrinsic contradiction between excessive centrality and efficiency in situations of rapid urban growth. When all distribution occurs through one or a limited number of points, a chain of events is inevitably set in motion. Initially, centrality generates rapid growth. Because the favoured point is by far the most desirable location, distributors and suppliers seek it out and the induced growth increases its centrality and its desirability.

Over time, however, problems inevitably set in. One of the most serious is traffic congestion. Food wholesaling generates an enormous amount of movement, whether by road, rail, or water. As congestion increases, severe cost increases are experienced by both distributors and suppliers. Additionally, extensive traffic management problems result, to the general detriment of the city. Ameliatory measures are usually introduced but these, almost inevitably, are short-lived: in the longer term they generate further problems. Thus, for example, in Bangkok, food-delivery trucks are barred from moving within the city limits between 0600 and 2100 hours, except on Sundays. Those trucks which are trapped by the deadline while delivering have no option but to park between these hours. This in turn places enormous demands on parking in and around the market. Parking for consumers becomes extremely difficult and this increases inconvenience and operating costs. A survey showed that the average time spent seeking parking around the Pak Klong Talad–Yod Piman market complex was 13 minutes and parking was frequently found only considerable distances away (Kuhn 1972).

Simultaneously, problems are inevitably experienced within the market itself. The lack of alternative locations increases pressure on the market: either congestion increases or the market is increasingly dominated by a limited number of dealers. Either way, costs increase.

Ultimately, relocation is demanded, but in the search for alternative sites, contradictory needs emerge. Smaller purchasers, such as informal vendors, who have little or no access to private vehicular transport, require ease of pedestrian or public-transport access. Larger consumers and suppliers require ease of vehicular access. Usually, therefore, relocation involves a move from places of maximum geographic (and economic) centrality to places of easier *vehicular* access, which are highly eccentric to smaller, poorer traders.

Significantly, it is extremely difficult to decentralize by moving only part of the market after excessive pressure has developed, because of the economic interconnections which emerge within the market over time, and because of political ferment around personal rights and who should move where. The clear implication is that a balanced, decentralized wholesaling system must be planned in conjunction with urban growth. Wholesaling facilities should be seen as an essential part of urban infrastructure in rapidly growing urban contexts.

Given the seemingly powerful case which can be made for a decentralized wholesaling system, it is important to evaluate critically the reasons usually advanced for highly centralized systems.

Arguments used to support a centralized wholesaling system

The first is efficiency: it is argued that centralized wholesale markets enable the farmer or farmers' agents to supply urban consumers most efficiently by providing a single drop-off point. Similarly, it enables consumers to purchase a full range of products in one trip. Clearly, the issue is scale-related. From the perspective of the farmer, the argument holds when the urban centre is fairly small. With increasing lateral spread of the city, however, an increasing number of farmers become further and further removed from the market. A point is reached when it is in the farmer's interest to be able to supply a more local market, and when the size of the local market is sufficient to absorb the majority of the farmers' output.

Similarly, from the perspective of the consumer using the market, increasing inefficiencies result from centralization with increasing urban growth. Perhaps the most important point, however, is that there is a qualitative dimension to the issue of efficiency. Centralized systems are fairly efficient for very large producers and consumers. With decreasing size of business enterprise, however, transportation costs assume a relatively increasing proportion of total costs. Excessive centralization therefore results in a situation in which both very small-scale producers and potential market-users are effectively excluded from the system. However, in urban contexts which are characterized by high levels of poverty and unemployment, arguably the overriding social and economic imperative must be to create opportunities for small economic enterprises.

It is important to emphasize, too, that the issue of spatial decentralization is inextricably interrelated with that of urban population density. In situations of very low density, centralized systems are the only ones which can work, since the lateral extent of the local market which contains the necessary threshold to support a decentralized wholesaling facility is too large for the cost equation to be altered significantly. The corollary of this is that the spatial centralization demanded by low densities promotes economic centralization and the dominance of large production and distribution elements within the urban economy.

The issue of decentralization of the wholesaling system has regional as well as urban dimensions. Thailand provides an example of this. The Pak Klong Talad–Yod Piman fruit and vegetable market in Bangkok plays an increasing role as transhipment point for the entire country, since regional wholesaling facilities do not exist. The consequences are severe.

> It is not in the least unusual that produce which is being transported to Bangkok is being hauled back more than half the way to its point of origin to reach the consumer. Of the total quantity of fresh fruit and vegetables received at the Bangkok wholesale markets, about 35% and 46% respectively is being re-shipped to up-country destinations.
>
> (Kuhn 1972: 11).

The second argument advanced for a centralized system is related to a price determination and co-ordination function. By centralizing wholesaling, it is argued, demand and supply can be co-ordinated better, prices set, and price fluctuations reduced. It is difficult to see why this function requires centralization. The argument for the decentralization of wholesaling functions relates primarily to basic foodstuffs which are locally produced (for example, fruit and vegetables). Products which are transhipped nationally or regionally usually enter the city at a fixed point and inevitably a degree of centralization will exist in relation to these. Many basic foodstuffs, however, are produced relatively ubiquitously in and around the city: similar products are grown in proximity to different segments of the city.

To the degree that local supply matches local demand, there will be considerable price stability. In the event that production exceeds or fails to meet local demand, prices will fluctuate. In this event, there is no intrinsic argument for a constant price of produce across the city: even with a centralized system real prices to consumers (as opposed simply to market-point prices) fluctuate in response to different transportation costs. Further, continuous electronic communications between wholesale markets are easily established. In the event of large surpluses occurring in any one part of the city and shortages in others, any one, or combinations, of three sets of actions can be set in motion: consumers in urban segments experiencing shortages can be directed to places of surplus; producers in segments of surplus can be directed to places of shortage; or produce can be transhipped via the market organization between one market and another. In any case, the system operates more efficiently in reducing total transport costs than one in which everyone is forced to move to a single point. For non-standard products, variations in specialization would inevitably occur, giving areas of the city different environmental 'flavours', with entirely positive consequences.

From the perspective of the urban consumer and the distributing vendors, the wholesaling system operates best when a wide range of products (for example, fruit, vegetables, spices, fish, meat, and poultry) are wholesaled in close proximity to each other. In practice, this frequently takes the form of different products making up different sections of the same market – a situation that can be quite satisfactory, provided potential incompatibilities between products (for example, fruit and vegetables, and fish) are resolved through market structure and sensitive design. In other circumstances, however, a system of wholesale markets can operate subregionally, with individual markets specializing in particular products but with close interconnections maintained between them (an example of this can be found in the Pettah district of Colombo, Sri Lanka). The advantage of this is that it tends to promote a rich pattern of relatively specialized local urban retail activity, without the degree of specialization being exclusive. Thus, distinct local flavours and characters (fruit districts, fish districts, and so on) tend to emerge. Decisions about whether to integrate in one market or specialize relate primarily to the volume of the activity (the smaller the scale, the stronger the argument for integration into a single market) and issues of traffic circulation and loading.

In summary, it is apparent that wholesaling markets represent a powerful element of urban structure. They generate considerable amounts of traffic; they frequently attract a wide variety of other activities (for example, eating outlets, retail markets, places of selling or hiring trolleys and other forms of conveyancing, suppliers of agricultural equipment and services, all of which seek these locations to be easily accessible to farmers unloading produce, and so on); they significantly affect the nature and costs of the food distribution system within cities and affect the number of people who may enter that system; and they affect the structure and the costs of the food production system. Clearly, therefore, their form and location need to be carefully examined from a synoptic urban structural perspective: they cannot be considered in isolation from broader issues of structure and form.

Factors affecting the location of wholesale markets

The location of wholesale markets should be informed primarily by five sets of relationships. The first is the relationship to agricultural producers. The aim here should be to ensure that all suppliers (especially small local suppliers) have easy access to marketing outlets. The second is the relationship to the local urban population which the market seeks to serve. The aim here is to enable easy access by consumers and small-scale vendors: appropriately, wholesale markets should be seen as feeders supporting a system of local markets and other forms of retail distribution. The third is the relationship to regional and city-wide transportation systems, particularly public transit, to ensure that access is not restricted by transport costs. The fourth is the relationship to other forms of wholesale outlets. Various forms of wholesaling should be seen as, and integrated into, a cohesive system to ensure easy subregional access to a full range of products. The fifth is the relationship to other large-scale urban elements. While the greatest possible integration of wholesaling outlets into the urban fabric is desirable, it is necessary to recognize that a pervading characteristic of wholesale markets is that they generate considerable amounts of movement, frequently involving heavy vehicles. When markets are located in such a way that the traffic generated by them intersects that caused by other large traffic generators, severe congestion, with resulting inefficiencies and increased costs, may well result.

Internally, the layout of wholesale markets can take many forms. Arguably, the most important factor determining this is efficiency of internal circulation. Internal vehicular circulation is usually required to all outlets to facilitate loading and unloading, and the most common form of internal inefficiency (with resultant cost implications) is caused by bottlenecks at access and egress points.

Administration and management

There are several issues relating to market administration and management which emerge from the case studies.

Responsibility

The first is who, appropriately, should be responsible for generating a markets system and for supplying market infrastructure: should it be undertaken by public agencies or by the private sector? Experience shows that it is not necessarily a question of either one or the other: in many contexts the two co-exist quite comfortably. It is important to stress, however, that the two forms of delivery agents have entirely different motives. In the case of the private sector, the motive is profit. By definition, therefore, for private markets to come about, conditions for profit generation via rentals must exist. The supportive population threshold must be large and levels of profit generation amongst vendors relatively high, since these markets inevitably have higher rental structures than public markets. Obviously, therefore, private markets tend to be dominated by larger, more affluent traders. The primary motive in terms of public market-provision is social benefit. If the intention of a markets policy is to generate employment, increase the circulation of income, and create maximum survival chances for all, including marginal operators, a significant proportion of markets provision must be public, and if necessary they must be subsidized – a large number of traders in developing countries cannot afford additional overheads, particularly in the early stages of their operations.

Fragmentation

The second issue of market administration is fragmentation. In many contexts, the administration of hawking and informal vending is fragmented between a number of local-authority departments (commonly licensing, health, traffic, and

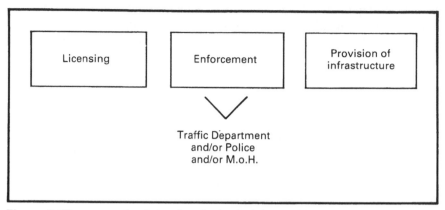

Figure 24 Fragmentation in local-authority decision-making structures

municipal police), without any one authority assuming overarching responsibility (Figure 24). A number of interrelated and negative consequences result from this. First, a cohesive policy seldom emerges. Hawking and informal vending are automatically treated as by-products of other areas of concern (for example, efficient traffic flow, crime prevention, and so on) and are by definition perceived from the outset as 'problems' affecting those other interests. Second, the agendas of different departments are frequently different and directly conflicting policies and attitudes may result. Third, the control dimension of policy automatically dominates. The public message which results is that hawking and informal vending are activities which, at best, should be reluctantly tolerated and a 'them and us' psychology is generated between traders and the local authority: traders see the authority not as an agent of assistance but as one of oppression, and mutual suspicion characterizes dealings between the two.

A successful policy demands that the *facilitative* and not the control dimension of policy dominate and that the authority is perceived by traders to have their interests at heart. This, in turn, demands that one section of the city administration be charged with giving form to policy and with creating maximum opportunities for small traders. Since, as has been argued, the creation of economic opportunities cannot be viewed in isolation from issues of city structure, form, and growth, it is appropriate that such a section be located within the urban planning department.

This does not mean that sections such as traffic or health no longer have any say about informal vending: the vending section needs to co-ordinate closely with other affected sections. It does mean, however, that when conflicts between different interests occur, they do not automatically result in punitive actions against traders. The conflicts are referred back to the vending department, and resolutions are sought and negotiated.

There are several advantages which result from this form of administrative structuring. First, the facilitative and the control dimensions of policy are separated. Conflicts in the field are still identified by agents of other sections (for example, traffic officers, health officers and so on) but they do not have the right to act unilaterally in support of the interests they serve. They report problems, and procedures are set in motion to seek conflict resolution. Once the authority is seen to be concerned about helping, as opposed simply to controlling, mutual confidence between traders and the city increases, two-way communication and information increases, and negotiation – and thus compromise – become easier. Second, most tensions between informal traders and city administrations result from the attitudes of officials in the field, many of whom regard informal-sector activity as a highly undesirable phenomenon. It is naïve to believe that these attitudes will change by themselves, from the bottom up. It is necessary to have a facilitative policy direction clearly established by one administrative section and that direction scrupulously enforced. Third, the issue of actively and creatively seeking locational opportunities for small operators becomes an integral part of city management.

Licensing

The third issue is whether a system of licensing traders should be enforced or initiated. Many cities in developing countries have such systems. Two motives generally underpin these. The first is revenue generation and the second is control – that is, licences are used to limit the number of vendors. Both need to be examined in greater detail. As an instrument of revenue generation, it is notably unsuccessful. The actual amounts raised are usually small and are frequently exceeded by the costs of collection and monitoring. Further, in contexts characterized by high levels of poverty and unemployment, there is no inherent logic in taxing attempts to generate income before any such generation occurs.

As an instrument of control, too, it is highly problematic. First, it leads to defaulting and thus creates an artificial division between legal and illegal operators. Second, it promotes insecurity amongst 'illegals' and this affects their trading environment in a negative way. As stated previously, security is a pre-condition for expansion, for improvements to market infrastructure, for control over littering and garbage dumping, and so on. Third, it is a non-progressive form of taxation in that it differentially affects smaller, more marginal traders – precisely those who have the greatest need for assistance. For these reasons, therefore, there is little point in licensing informal traders.

Levels of infrastructure

The fourth issue is the definition of appropriate levels of market infrastructure. In most countries local-authority investment in market infrastructure is usually geared towards improving the trading environment (for example, by providing shelter to enable a market to operate successfully in all types of weather) and promoting hygiene (for example, by providing cleanable floors and selling surfaces, drainage, water, garbage removal, etc.). There is, by definition, no single appropriate level of infrastructural provision: it varies with contextually specific factors such as climate, products sold, urban location, scale, affordability of traders, and so on. The following general observations are, however, apposite.

First, the level of infrastructure affects market costs. Nowhere do markets conducted in formal market-buildings, with a full range of services, pay for themselves. Second, while the principle of cost recovery should inform policy, some form of subsidy is usually involved. Urban markets, therefore, should not be seen as revenue generators but as investments in social infrastructure. Third, many informal operators have extremely low entry ceilings. Unrestrained attempts to recover costs may well price out a great many traders and it is always the most marginal ones who will be affected the most. The principle of allowing as many traders as possible to operate should define the level of subsidy. Fourth, there is no close correlation between levels of infrastructural investment and the success of markets. As a general principle, investment should be kept as low as possible, relative to achieving social and environmental objectives which are appropriate to the context.

Issues relating to the administration of individual markets

Stall allocation

An important factor affecting the performance of markets is the way in which market entry and stall allocation occur. There are several issues which are germane to this.

The first is the question of who should control entry – should there be external regulation via the market authority or internal regulation via market operators? As a general principle, the market authority should retain control. There are frequently tendencies towards oligopolization within urban markets and cartels of larger traders may well emerge. It is essential, therefore, that the authority imposes the principle of equity of entry to the greatest possible extent.

The second is the time period of stall allocation and the means of allocation. The experiences of other countries provide a number of pointers in this regard. One is that allocation of stalls should not be permanent or even made for extended periods of time. Although long-term allocation approaches may be administratively easier, they are extremely inequitable and gross distortions in social purposes may result. The squatters' market in Durban, South Africa, provides an example of this. Rights to stalls in this highly successful market were auctioned some years ago: operators have ongoing rights provided a nominal monthly rental (approximately £22·74) is met. If an operator ceases to trade from the market he or she may transfer rights. While the council makes a nominal entry charge (of four times monthly rental) for this, the goodwill trades for large sums of money. It is, of course, impossible for smaller traders to gain access to the market. Consequently, they attempt to take up 'illegal' interceptor locations outside the market. This is resisted by the permanent traders and the local authority is caught in the middle of considerable antagonism and conflict.

As a general principle, the shorter the time period of allocation of stalls, the more equitable the system. The actual length of stall allocation will, however, need to vary with the degree of permanence of stall-holders and the size of the individual business. Larger, more permanent traders may need to invest in stall infrastructure, but will need a degree of security of stall tenure if they are to do this. Smaller daily traders who need little additional fixed stall infrastructure should be allocated stalls for shorter periods of time, and periodic or temporary traders may need a system based on the sale of daily (or even half-daily) tickets. The central fresh produce market in Harare, Zimbabwe, for example, is divided into a number of different sections. In certain parts of the market larger traders are allocated stalls on a longer-term basis and in other sections traders purchase a daily ticket for Z$2·00 (90 pence), which gives them access to a space marked out on the ground.

Related to this issue are procedures of stall allocation. Three systems are commonly applied: auction or tender; ballot draws relating to particular stalls; and allocation on a first-come, first-served basis. The first pushes up stall prices and, almost by definition, excludes smaller traders. The second makes it extremely difficult to maintain product specialization and the haphazard mixture of uses which result may detrimentally affect market performance in general and differentially affects particular traders negatively. The most equitable system pertains when allocation occurs on a first-come, first-served basis and when rentals are low enough to allow entry to all.

The third issue is the spatial dimension of allocation. Central to this are concerns about marginalization and product specialization. As discussed, it is important to maintain a high degree of product specialization. When allocation occurs on a first-come, first-served basis, and particularly when there is a high degree of periodicity in trading patterns, this may be difficult. The best situations result when traders are classified by use and when allocation occurs in terms of use zones. Rather than conceptualizing markets in which a variety of products is traded as single entities, therefore, they should be viewed as interdependent systems of sub-markets comprising different uses.

Similarly, when there is a high degree of periodicity, spatial marginalization may occur as the market contracts in non-peak periods. Permanent traders may be left isolated as periodic traders leave and may effectively be excluded from consumer search patterns. Two practices can alleviate this. The first is initially to classify traders into permanent and periodic categories and, as far as is possible, to allocate permanent traders around the market core, with periodic traders coming and going (within use zones) on the periphery of this. The market would thus contain a permanent core and a periodic section and would be made up of a system of interlocking linear elements (which could be integrated into a variety of forms) which maintain use specialization.

Controlling tendencies towards economic dominance and monopolization

A feature of most urban markets, even those involving small traders, is the tendency for larger traders to seek to dominate smaller, more marginal ones to their own advantage. This tendency must be monitored continually and checked.

There are four types of actions which are important in this regard. The first, which has already been mentioned, is preventing organizational cartels of traders from regulating and determining market entry. The second is the prevention of excessive physical expansion of any one trader. Once markets have been divided into basic selling areas or modules, therefore, there must be an upper limit on the number of modules which any one trader can occupy. This is normally restricted to two or three. The third is grapevine monitoring and, if necessary, expulsion. A common form of take-over is for larger traders within the market, or even for formal economic enterprises from elsewhere, to seed the market with 'fronts' or 'runners', who operate as branch enterprises for the parent operation. The only way in which this can be detected is through monitoring and keeping an ear to the ground. A fourth type of action relates to maximizing the amount of well-located space available in a market. Where the number of desirable locations in a market is limited, the stronger, more powerful operators will attempt to gain control of them. Market size and market design therefore become important tools for achieving the social purpose of a markets policy.

A certain amount of expansionism and monopolization may not, depending on circumstances, be a major problem. However, as soon as it becomes excessive it must be controlled if smaller, more marginal operators are to have a chance and if genuine consumer choice is to be promoted.

Promotion of market organization

Frequently, once markets have become established, a formal organization of market traders emerges. Such organizations, if democratically based, are a great advantage: they facilitate two-way dialogue and become a conduit whereby the concerns of traders can be transmitted to urban authorities and vice versa; they can negotiate agreements with members about market behaviour and garbage removal; they can assist in policing and action against petty crime; and they are invaluable as a mechanism through which additional back-up services (for example, collective bulk-buying, vehicle pools, shared refrigeration, savings and credit unions, and so on) can be initiated. For these organizations to be positive, however, they must be genuinely democratic. Where democratic organizations are slow to emerge, there is a strong case for the market authority facilitating the process. There are three main ways in which this can occur: by indicating a clear willingness to initiate dialogue, discuss, listen, and act; by showing a preparedness to devolve functions and decisions which could appropriately be taken by the traders; and by supplying or organizing seed-finance for elementary administration and for back-up assistance.

Chapter 3

The empirical foundation

The nature of informal selling in the seven case study areas

This first section provides a general description of informal trading in the seven cities, with regard to its extent, its degree of formality, location, and the degree of specialization within markets.

Extent of informal trading

The numbers of informal traders in any city may primarily be regarded as a product of three factors: the extent of un- and underemployment and poverty (which will force people to attempt to survive through informal activities); the extent to which the urban system creates trading opportunities to which they can respond; and the effectiveness of controls imposed by the authorities in relation to those opportunities.

In terms of the first factor, the seven cases considered here range from Bombay, at one end of the spectrum, where unemployment and poverty are very high and effective control is minimal, to Singapore at the other end, where unemployment is much lower and controls are exceedingly strict. In Bombay informal traders are to be found in almost every street and open space, with most depending for their survival on the few rupees a day which they can earn from trading. In Singapore, by contrast, street sellers have been confined to the few alleys and backstreets set aside for them. All other informal traders have been removed to formal market-buildings.

Most other cities (in terms of the extent of informal trading) fall somewhere in between these two extremes. As may be expected, there appears to be a correlation between the level of socio-economic development of a country and the size of the informal sector. Relatively large numbers of street traders are to be found in Harare, Colombo, and Bangkok, and somewhat fewer in Taipei, Hong Kong, and Singapore.

Degree of formalization of hawking

To a large extent, this depends on the policy adopted by particular authorities towards the provision of market infrastructure. Bombay has a large number (ninety-two) of formal fresh-produce market buildings but there are no large-scale projects underway to build additional ones, or to provide any permanent infrastructure for the many thousands of street traders. As a result, most informal trading takes place in on-street locations, using makeshift infrastructure which is generally provided by the seller. Much use is made of the existing environment: thus, trees provide shelter, the tarred road a selling surface, and for many street traders their selling surface doubles as a bed at night. There is also a high incidence of selling from small kiosks (commonly 2·5 by 4 metres in size) set into the ground floor of on-street building façades.

One reason for the general lack of more permanent street-market infrastructure undoubtedly has to do with insecurity of the traders themselves. Many are 'illegal' in the eyes of the local authority and are subject to frequent harassment by hawker patrols. Traders who have to disappear whenever the hawker patrol approaches are unlikely to spend time or money on providing stall infrastructure.

In most other cities greater attempts have been made by the authorities to formalize street trading, either by designating areas from which it can operate or by providing infrastructure in the form of shelter, surfaces, taps, bins, toilets, and so on. Often, attempts by the authorities to remove hawkers from central business districts, and to 'neaten up' street selling, have been linked to the public provision of infrastructure in locations considered more 'appropriate' by government. Frequently, these alternative locations have been entirely inappropriate for street sellers and stand unused as a result.

In Colombo, permanent street-market stalls have been widely provided in the busy central (Pettah) area: some are well located and are intensively used but in other cases stalls lie abandoned. Outside of the Pettah area market buildings have been provided, but they are generally small and much street trading takes place outside them. Less effort has been made by the authorities outside the Pettah area to provide street-market infrastructure.

In Bangkok, the system of official nucleated markets is well developed, and there are many very large markets. New fresh-produce markets have been built outside of the old central city as urban expansion has occurred. In Bangkok, as elsewhere, it is common to find informal street markets established adjacent to formal market-buildings. From observation, it would appear that the two usually operate harmoniously together, with sellers in the informal retail market selling at somewhat different times of day and in smaller quantities than those in the formal market. In effect, the formal market frequently plays the role of a second-level wholesaling outlet to the informal market. Pavement selling is widespread in Bangkok but is far more ordered than in Bombay, and hygiene and cleanliness appear to be less of a problem. Part of the reason for this is that general standards of infrastructure and services are higher than in, for example,

Bombay: pavements are generally hardened, drainage is good, street cleaning is effective, and so on. Further, the local authorities appear to have played some role in regulating street markets, designating suitable locations and providing cleaning services. Stall infrastructure is generally provided by the stallholder: canvas or umbrella shelters and selling surfaces made from baskets and boxes are cheap, effective, and usually promote an environmental condition which is conducive to trading.

Hong Kong is actively pursuing a policy of providing formal market buildings: this involves both the renovation of inner-city market buildings and the construction of new ones in housing estates. Markets are regarded as an essential part of the infrastructure of housing estates. Although the existing policy is eventually to move all street traders into these formal market-buildings, extensive street trading still exists in Hong Kong. As in Bangkok, it is relatively well regulated and serviced, with the bulk of stall infrastructure being provided by stall-holders themselves. Again, this infrastructure consists primarily of umbrellas and canvas for shelter and boxes or baskets for selling surfaces. A common, more formal, form of provision consists of municipally owned and rented corrugated-iron kiosks which can be closed up and locked at night. Numerous streets are lined on both sides with these kiosks. When open during trading hours, they create successful and environmentally positive linear markets; when closed, however, they create a sense of sterility.

Similarly, in Taipei, the authorities are attempting to move street traders into formal market-buildings, but there, as elsewhere, extensive street trading still takes place. Street selling occurs under a variety of conditions, ranging from canvas and umbrella cover to more permanent on-street kiosks and nucleated markets under low-cost roofing. A very successful example of the latter is the clothing and cooked food bazaar opposite the Longshun Temple in central Taipei. This market, measuring some 6,900 square metres in extent, is covered by perspex roofing on iron-pole supports. Stalls consists of lock-up kiosks (± 2 by 2 metres) with tilt-door fronts and are arranged in a grid-iron pattern. The market specializes in clothing and household goods and there is a large sit-down cooked-food section. The market succeeds in providing diverse selling conditions in a cheap but highly attractive environment.

Taipei also has a number of informal night markets. Certain streets are closed to traffic after 2000 hours, and selling kiosks are rapidly assembled from bamboo poles and perspex covering. Electricity connections are made to adjacent formal shops (for a fee) and the environmental effect is dramatic. The scale of these markets and the degree of specialization are considerable: for example, one such market, selling clothing alone, consists of a double-sided run of some 130 metres; another specializes in cooked food and extends over 230 metres down either side of the street.

In Harare, the official attitude to informal trading is that it is a symptom of underdevelopment and consequently that it should be stamped out. Tempering this is a more pragmatic realization that, given high levels of poverty and unemployment, it cannot be removed entirely, and that it should thus be treated as a necessary evil. The policy outcome of these somewhat contradictory perceptions is strict enforcement action against informal activity in the central business district (CBD) and other high-profile areas (such as the more up-market shopping centres or higher-income residential areas) but a far more relaxed attitude in outlying low-income areas. Additionally, a policy of providing semi-formal 'People's Markets' (usually consisting of open-sided, covered, permanent structures with water points and permanent raised selling surfaces) has been introduced. Some twenty-one of these are planned: to date, four have been built and a further six have received development approval. In the outlying residential areas, considerable non-regulated trading occurs, particularly along arterial roads and bus-stops or terminals. The form of selling infrastructure varies considerably, from displaying on the ground to the erection of semi-permanent kiosks, usually of wood and corrugated iron.

Singapore presents a different picture from most other cities covered here. The removal of street traders into formal market-buildings with high levels of infrastructure has been proceeding rapidly. Remaining street traders have been collected into designated streets and provided with basic services. In these situations, traders have generally provided their own stall infrastructure, which ranges from a basic selling surface made from boxes or baskets, to wooden kiosks which can be locked at night.

The greatest concentration of street traders is to be found in the old Chinatown area, on the edge of the CBD. Trading occurs from wooden surfaces and kiosks or from shelters of canvas or umbrellas, and these line the streets and fill the public squares. At night, the streets are closed off and a night market is established. A stall framework of bamboo poles is erected and covered with canvas, neon tubes are tied on, and electricity connections made from nearby shops and houses. This highly successful market is nearing its end, however: a large, modern concrete structure is being constructed nearby and it is the intention to move all street traders into this. In the process, undoubtedly, the area will lose much of its colour and vibrancy, and many smaller street traders will suffer economically through the move.

The issue of the level of infrastructure provided for small-scale traders is a crucial one. Clearly, a range of infrastructure is necessary to cater for the extensive differentiation which occurs within the category of informal trading: larger traders may require, and can afford to pay for, higher levels of infrastructure and service; others can afford very little, if anything. A frequently observed situation in the case studies covered here is one in which small traders are being forced either to move into expensive and elaborate market buildings, or to trade illegally from street locations. For many of these sellers, the latter is the only alternative and problems are consequently created both for the city authorities and for the trader.

Location of informal trading

It is necessary to distinguish here between patterns of hawking which were established spontaneously, and those which emerged as a result of local authority intervention and planning. Where location has been relatively spontaneous, there are three major factors which influence the pattern of trading.

The first factor is the location of generators of population movement, such as railway stations, bus-stops and termini, boat and ferry loading points, and so on. The importance of these generators can be observed in a number of cases. In Bombay the largest rail terminus (Victoria Station) is located in the central city, and is in close proximity to the largest wholesale market (Crawford market): it is here that the largest concentration of street trading occurs. Markets line all the main routes from the station. In Colombo, the largest concentration of formal and informal markets is in the central Pettah area. Here, the centrally located main bus and train termini generate large population flows which in turn provide custom for the numerous street markets, the fresh-produce retail markets, and the fish- and fresh-produce wholesale markets. Outside the Pettah area, formal market-buildings have been built in close proximity to the main railway stations and there are, in most cases, open-air markets adjacent to these. In Hong Kong, a great deal of population movement takes place between the islands. Ferry landing points are therefore major population receptors, and all forms of commercial activity, including informal trading, tend to cluster around them. Finally, in Harare, the main Musika wholesale and retail market is in close proximity to the major regional bus terminus.

A second element informing the location of hawking is the alignment of the major movement routes: these may be viewed as 'linear generators' as opposed to the 'nodal generators' described above. Movement routes channel pedestrian and traffic flow, and form ideal locations for hawkers: they enable the traders to locate themselves in direct proximity to potential customers. Concentrations of sellers are therefore usually found along roads leading from major transport termini (Bombay, Colombo, Hong Kong), along central spines of commercial activity (Bombay, Hong Kong, Bangkok), or as in Bangkok, on some of the bridges which cross the klongs (canals). In Bangkok, the klong system is a major movement system and all the larger wholesale markets have located on them. Some 28–38 per cent of all vegetables and 30–50 per cent of all fruit is delivered to these wholesale markets by boat from small farmers on the outskirts of the city.

A third factor commonly affecting patterns of hawking is the distribution and density of residential activity in a city. In cities such as Bombay, Bangkok, Hong Kong, and Taipei, there is a high degree of land-use mix (particularly commercial and residential use) and high population densities, spread more or less evenly across the city. The result of this is that the intensity of activity is high, and hawking, while reaching points of greater intensity near nodal or linear population generators, is found ubiquitously across the city. In cities such as Colombo, Singapore, and Harare, where residential densities are lower and where there is a greater degree of separation between residential and commercial areas, the location of small traders is influenced to a greater degree by the presence of nodal and linear generators. In Singapore, for example, much of the older mixed commercial and residential area has been destroyed through urban-renewal programmes: however, in the only section of this kind left (old Chinatown), there is a high intensity of street-market activity.

In those cities where the building of new formal markets has been an important part of government policy (for example in Hong Kong, Singapore, and Taipei) the locational pattern of these markets is significantly different. In both Hong Kong and Singapore it has been accepted policy to relocate clusters of street traders in the central city area into formal markets close to their original location. Outside the central-city area, however, formal markets have been built as part of new suburban 'neighbourhood' developments. In Taipei, by contrast, it has been government policy to spread new market buildings evenly across the city in order to maximize their accessibility.

The location of street trading is thus determined by patterns of accessibility: the accessibility of traders to potential customers; the accessibility of consumers, and particularly low-income and therefore less-mobile consumers, to the goods provided by street markets; and the accessibility of retail fresh-produce markets to their source of supply – the wholesale fresh-produce markets. A high degree of accessibility is achieved when a high-density mix of commercial and residential land-use extends across the city. Under such conditions, high population densities provide a ready customer pool for small traders and most streets and public spaces provide a locational opportunity for them. This situation works to the benefit of consumers as well: purchasing selection and convenience is maximized. In a number of the cases covered here (Bombay, Colombo, Bangkok, and Hong Kong) a high-density mix of commercial and residential land-use is to be found in the older central parts of the city, and for historical reasons the main wholesale markets are also located here. Such conditions are particularly favourable for hawkers, as they have cheap and rapid access to wholesale suppliers of fresh produce.

The situation is somewhat different where markets have been built in planned 'neighbourhood' developments. In both Singapore and Hong Kong, for example, markets are usually provided as part of a shopping centre, which is internal to that neighbourhood. This pattern of development is negative in a number of respects. First, while the residents of these planned neighbourhoods theoretically have easy access to commercial and market facilities, distances are in fact such that they can often be reached only by car or bus. This discriminates particularly against lower-income, less-mobile people for whom fresh-produce retail markets are an important source of cheap supplies. Second, residents of a neighbourhood development find it very difficult to make use of market facilities in *other* neighbourhood developments: there is little mutual reinforcement of commercial activity. The distance between developments discourages this, and the commercial centres of different developments are often not well connected by public transportation. As a result, residents' choice of outlets is restricted. Since these markets serve a high proportion of captive consumers (who have little or no alternative outlets), stall-holders in these complexes are frequently in a quasi-monopolistic position and are therefore under less pressure to maintain competitive prices and quality of goods. A third effect of this pattern of development is that small traders find themselves tied to a customer pool which is highly periodic in terms of spending patterns: where most family members work outside of the area, buying is confined to early morning or evening and traders have little custom in the intervening hours.

Specialization of market activity

In most cities there is little *areal* specialization in terms of the kinds of goods sold in markets: it is usual to find a wide range of goods being sold over relatively short distances. Bangkok is somewhat of an exception to this and the fresh-produce wholesaling sector reflects some geographic product specialization. It would seem that factors of market size and ease of interconnection (by water and road) have allowed this specialization to take place. Where thresholds (i.e. the intensity of potential customers) are lower and where mobility is lower (either because of poor transport connections or consumer poverty), a more diverse and integrated produce pattern is more appropriate.

While most cities covered here do not demonstrate areal specialization, they do reflect considerable market specialization: particular markets tend to concentrate on the sale of particular goods. Thus, in most cases, it is possible to find specialist fruit and vegetable markets, fish markets, meat markets, clothing markets, household-goods markets, spice markets, cooked-food markets, and so on. Again, the degree to which this specialization occurs appears to depend on threshold: a high density of population allows for greater specialization. In Colombo, for example, the busy central Pettah area has highly specialized markets in both the wholesale and retail sectors. In the quieter residential areas outside Pettah, individual markets sell a wider range of goods. Generally, the planned markets of Colombo, Singapore, Hong Kong, and Taipei also show a relatively low degree of specialization. Most of these markets are designed to sell a fixed range of products: meat, fish, fruit and vegetables, household goods, clothes, and cooked food.

From the consumer's point of view, it is clearly preferable to have a wide range of products available within relatively easy reach. However, it is also important that a sufficiently wide range and choice of each good is available. The best situations therefore obtain when different specialist markets occur in close proximity to each other (such as is found in Bombay, the Pettah area of Colombo, Bangkok, Hong Kong, and in the central area of Taipei) or when non-specialized markets are large enough to provide adequate choice in relation to any particular product.

The management and administration of markets in the seven case-study areas: an evaluation

Approaches to the management and administration of hawkers and markets vary across the case-study areas. In no case can it be said that the management system is functioning optimally and there are no systems which present themselves as a 'role model'. Nonetheless, some important lessons can be learned from observing the problems experienced in the different areas. The systems are described in relation to the general attitude of officialdom to the informal sector; the structure of the administrative system and the allocation of responsibility for street trading and markets; the forms of control over hawking; policies adopted with regard to the provision of market infrastructure; and approaches to the issue of stall allocation in markets.

Official attitudes towards informal-sector selling

It is important to distinguish between officially adopted attitudes to informal selling and the implementation of policy. In all the cities reviewed here, a generally negative attitude towards informal selling (and particularly street trading) was expressed by the authorities, but the extent to which control is enforced varies widely from city to city.

In terms of official policy, all city authorities held that informal selling should be contained, or, if possible, phased out. This feeling was expressed *less* strongly by the officials in Harare (who, while viewing it as an indication of underdevelopment, appeared to accept that it was a necessary means of income generation for the poor and unemployed), by city officials of Bombay (who are acutely aware of the impossibility of such a task), and by the Colombo authorities. The Taipei authorities indicated an understanding of the factors giving rise to informal selling (small traders have evidently increased dramatically over the last few years as a result of economic recession and growing unemployment) but nonetheless expressed determination to relocate most street traders in formal market-buildings.

A control-orientated policy towards informal sellers was expressed *most* strongly by the Singapore authorities, and it appears that this attitude has its roots in a political motivation. When the People's Action Party came to power in 1965, at the time of independence, it was estimated that there were some 50,000 hawkers in the streets of Singapore. Moreover, they were highly organized and represented a strong lobby in opposition to the government. The organization was subsequently banned and the very strict controls and phased removals of street sellers have been interpreted as an attempt to break the strength of the hawker organization. In addition to this, there has been a shortage of labour in Singapore, and authorities have been attempting to draw street sellers into the formal job market. Hawking has been defined as 'unproductive labour', and to qualify for a licence a hawker has to be over 35 years of age or disabled.

In terms of the practice of control over informal trading, city authorities can be ranged on a continuum: Bombay authorities at one extreme have virtually no control over hawking; Singapore authorities, at the other extreme, have very effective control.

In Bombay, the situation is virtually anarchic: the scale of informal selling is such that control is almost impossible. Sporadic attempts to control street trading are evident, in terms of hawker patrol vans, but no sooner have these passed than traders are back at their posts. Reports of bribery and corruption amongst hawker police are widespread.

In other cities, attempts at control have been partially effective. In Taipei, one source reported that bribery and corruption amongst police (who implement hawker control) was a major reason for the lack of effective control. In Colombo, Bangkok, Hong Kong, and Harare, control measures have usually been more strictly applied, especially in the office and commercial centres of the city, but outside these areas enforcement is intermittent. The conflictive attitude expressed by officials in Harare (that it is undesirable but inevitable and, perhaps, even necessary) has resulted in an inconsistent policy approach: control is strictly enforced in the CBD and other 'high profile' areas; no control exists in the lower-income residential areas; and a 'supportive' policy of building 'people's markets' is being implemented.

The Singapore authorities have a 'three phase' policy for the removal of street trading. The first phase consisted of moving street traders from main streets to back streets and alleys and providing basic market-services for them. This phase is complete. The second, and current, phase consists of the building of nucleated market buildings in both the city centre and the suburbs and moving street traders into them. The first of these markets was built in 1972 and it is estimated that by 1989 all street traders will be re-sited. Thereafter, in the third phase, authorities will concentrate on policing the new system and ensuring that no further informal markets emerge.

The reasons given by authorities for their control-dominated approach are relatively similar. Most authorities cite problems of traffic and pedestrian congestion; hygiene problems resulting from the selling of contaminated food, litter, the clogging of gutters and drains with waste vegetable matter and human excreta where public toilet facilities are inadequate; and problems experienced by shopkeepers when hawkers take up positions immediately outside their shops. One complaint by shopkeepers, although not an undisputed one, is that hawkers provide unfair competition as they have much lower overhead costs and can therefore sell at lower prices. Other formal traders, particularly larger ones, counter this argument, however, by pointing out that hawkers seldom retail the same products as shop owners and that the increase in consumer activity caused by a larger concentration of sellers is generally good for business. The most frequent complaint from shop owners is that hawkers physically prevent customers entering their shops by blocking entrances, or that they obscure display windows. In practice, shopkeepers have in many cases adapted to this problem: practices such as the illegal renting of pavement space outside shops, the selling of electricity to hawkers, the leasing of storage space, and even the provision – for a price – of 'escape holes' to enable hawkers to avoid police harassment, were all observed.

A major problem has emerged in those cities where there is ambiguity over the continuing existence of street sellers. On the one hand, because street traders are viewed as a negative and temporary phenomenon, little effort is made by the local authorities to provide for them. Many of the problems relating to litter, unsanitary conditions, and conflict between vehicular and pedestrian flows and hawkers, are primarily a result of the failure of local authorities to provide basic services such as litter bins, hardened surfaces, taps, and suitable locations for street trading. On the other hand, sporadic attempts are made to remove and control street traders, and traders find themselves subject to arrests, fines, and general harassment. As a result of this, many of them are highly insecure and make little effort to keep their areas clean, provide basic infrastructure for themselves, or sort out locational conflicts. These factors in turn reinforce official attitudes towards the traders.

Responsibility for hawkers and markets

In most cases, the management and administration of hawkers and markets is highly fragmented. Generally, there is a division between those departments connected with administration and registration, those concerned with enforcement of regulations, and those concerned with the provision of market or street-market infrastructure (Figure 24).

Bombay, Colombo, Bangkok, and Harare all have separate hawker licensing departments within the local authority, although in Colombo and Bangkok the function of this department is simply to register the number and names of hawkers, and not to charge licence fees. In Bombay, Colombo, Bangkok, Taipei, and Harare, enforcement of regulations falls under the local police or traffic department, and in three of these cases (Colombo, Bangkok, and Harare) the local medical officer of health also carries out inspections of food vendors. Finally, the provision of market infrastructure is usually carried out separately from the above two functions: in Bombay, this responsibility is split between the local markets administration (responsible for market buildings) and the Traffic Department (responsible for demarcating areas for street hawkers); in Colombo, provision of infrastructure falls under the local Department of Veterinary Services, primarily because the current policy is to combine meat, fish, and fresh-produce sales into single 'supermarkets'; in Taipei, infrastructural provision falls under the local Department of Economic Construction.

By contrast, two cities (Singapore and Hong Kong) have highly centralized forms of market and hawker administration. Singapore represents the most consolidated case: here, all functions pertaining to hawkers and markets fall under the central government Department of the Environment. This department registers and licenses, inspects and administers, and finances street and formal markets. Hong Kong has, to some extent, followed the Singapore model: all hawker and market functions fall under either the local Urban Services Department (responsible for commercial and industrial areas of the city) or the local Housing Department (responsible for housing estates).

These fragmented systems of management and administration usually reflect the absence of a coherent policy towards hawkers. Such systems have frequently evolved in an *ad hoc* manner as different problems and needs emerged; and, in general, they reflect the attitude on the part of many local authorities that street vending should be subordinated to the dictates of traffic and pedestrian flows, and should be replaced where possible by consolidated market buildings.

Fragmented administration systems give rise to a number of problems. First, the Traffic Department and its enforcement officers tend to play a dominant role with regard to the location and operation of street hawkers and this role is nearly always a negative and restrictive one. Those departments responsible for the provision of infrastructure for hawkers therefore find themselves subordinate to the dictates of traffic. Second, the responsibility for infrastructural provision for street traders often lies outside the scope of the various licensing, enforcement, or markets departments. Street trading therefore becomes an activity which is tolerated in certain places but rarely promoted. The possibility of a promotional markets policy involving the entire range of small operators – from the smallest and most marginal street trader to the larger, more established trader who may require premises in a formal market-building – is thereby negated. Third, the frequent absence of a formal decision-making linkage between the departments concerned with enforcement and those concerned with promotion and infrastructural provision means that problems relating to the operation of street traders are seldom resolved in a positive manner: the possibility of a market system being actively used to solve a problem of conflict between traffic and street traders is excluded by departmental fragmentation. Fourth, certain departmental functions have become relatively superfluous: the issuing of licences to hawkers, which in some cities has a charge attached to it and in other cities not, is in most cases entirely ineffective either as a control mechanism or as a registration system: often, 50 per cent or more of street traders are unlicensed and enforcement officers are able to make little impact on the situation.

The centralized administration systems of Singapore and Hong Kong have resulted in a more co-ordinated policy regarding hawkers and markets: the problem in both these cases, however, is that the *direction* of the policy is restrictive rather than promotional. In Singapore, the aim of the policy has been to move street hawkers into formal market structures and the authorities have been relatively successful in achieving this, albeit by very authoritarian measures. Similarly, in Hong Kong, markets falling under the Housing Department are well used, primarily because of the strict control on hawkers establishing themselves in more favourable and cheaper locations outside the formal market-buildings. Such an approach does not, however, foster job creation or the alleviation of poverty through the creation of survival opportunities for the small and intermittent trader.

Forms of control over hawking

Three main forms of control are to be found in the seven cities.

Licensing restrictions

In all cases, small traders are required to apply for licences, although in Harare this applies to on-street traders only: vendors trading from designated markets require a market-stall ticket only at the time they are trading. In two cases the licensing system is used to control the numbers and types of traders: in both Singapore and Hong Kong, no new licences are issued to street hawkers, and traders in on- and off-street locations have to be over a certain age (35 years in Singapore and 40 years in Hong Kong) to hold a licence. In Singapore an applicant for a hawker's licence also has to be certified as being of good character by a Member of Parliament. Hong Kong places restrictions on the types of goods which licensed hawkers can sell. Only dry goods are allowed to be sold in street locations, and the selling of food is restricted to certain defined locations.

In other cases, care is taken to ensure that the licensing charge does not become an exclusionary measure. In Bombay the licence charge is kept deliberately low (20 rupees [£1.00] a month in 1983) in order that the very poor are not priced out, and in Colombo and Bangkok, no charge is attached to the licence. Only in Hong Kong are both licence fees and fixed-pitch fees charged, and the subletting of space is forbidden. In Harare, the annual licensing fee was a nominal Z$2·00 (90 pence).

Restricted areas

All seven cities attempt to control the location of street traders by defining areas from which they may or may not operate.

In most cases, selling in the CBD of the city is severely restricted and attempts are made, instead, to define a limited number of zones or street where hawkers are allowed to operate. In Bombay, however, zones are designated along particular streets and pavements and certain streets are closed after 1600 hours to allow temporary markets to be set up. Bangkok has 269 areas designated for hawking. Further, certain streets are closed in the evenings or over the weekend, and on wider roads, temporary barriers are erected on Sundays, cutting off the lane closest to the pavement and creating space for a linear market. Hong Kong designates certain streets for hawker trading and defines, through road marking, sites from which hawkers are allowed to sell. Hawkers are barred from entry into housing estates by barriers placed at all the entrance roads. In Singapore, the only street markets which are tolerated are temporary ones from which traders are destined for removal to formal market buildings. Certain open areas or car parks have, however, been turned into open-air food centres, providing space for numerous small cooked-food sellers.

Enforcement and inspection

All the cities have some system of enforcement and inspection: in most cases, this lies in the hands of the traffic police and health inspectors. Generally, traffic police or municipal police are responsible for enforcing both licence and locational regulations outside built markets. The local medical officer of health usually also has an inspectorate which carries out checks on traders both inside and outside formal market-buildings. Singapore and Hong Kong are exceptions to this: in Singapore, officials of the Department of the Environment enforce hawker and market regulations, including health standards, and in Hong Kong the local Urban Services Department and the Housing Department each have their own inspectorate.

The ability of authorities to effect enforcement is obviously related to the quality of the inspection force and the powers at their disposal. In Bombay, Singapore, and Hong Kong, *de jure* powers are considerable. Officials can issue fines, confiscate hawker goods, and arrest offenders. In Singapore, the tickets issued for infringements serve as automatic summonses to a magistrate's court, although admission of guilt fines can be paid.

Perhaps the most important lesson to emerge from these cases is the fact that, if widespread poverty and unemployment exist, people are forced to pursue survival strategies, regardless of regulations. Under these circumstances, only the most draconian system of control and enforcement will keep hawkers off the streets and pavements in areas where there are potential customers. Excessive attempts to control and restrict street trading in cities where there is poverty and unemployment inevitably create conflict, and while they may succeed in making life more difficult for hawkers, they will not eradicate these activities.

Bombay, a particularly poor city, illustrates this point. The system of regulatory controls and enforcement appears to have little or no impact on numbers and locations of street hawkers. As the hawker patrol van appears at the end of a street, hawkers pick up their goods and disappear into neighbouring shops. Almost as soon as the back wheels pass their location, the traders begin the process of re-establishing themselves. The only effects are, first, that illegal hawkers generally carry a very small quantity of stock which they can easily move as a patrol van approaches; second, that illegal hawkers show less concern for cleanliness and rubbish removal in their environments, because of their often temporary nature in any one location; and third, tendencies towards bribery are increased. Such effects are negative, in terms of the ability of the hawker to generate profits, in terms of the cost to the city administration of additional cleaning, and in terms of the credibility of city government.

Even in a city such as Hong Kong, where powers of enforcement are more extensive than elsewhere, there is considerable illegal hawking. An important factor in some of these cases appears to be the quality of the enforcement officers:

often their pay is so low and conditions of work so bad that they are open to all kinds of bribery and corruption from illegal hawkers. Such a system also negatively affects the survival chances and profit generation of the smaller and weaker traders. Only in Singapore, which has relatively low unemployment, does there appear to be a considerable degree of official control over informal traders and this occurs at the cost of extensive policing.

The provision of market infrastructure

The way in which market infrastructure is provided and financed is strongly influenced by the overall policy towards market provision. In all seven case studies, it is accepted policy that market infrastructure should be publicly provided and in every case this infrastructure primarily occurs in nucleated, built markets. Thus, Bombay has ninety-two fresh-produce markets, and is building additional ones; Colombo is engaged in the construction of fresh-produce 'supermarkets'; Bangkok has a large number of fresh-produce markets, but is relying increasingly on the private sector to provide new ones; the Singapore authorities have constructed large numbers of fresh-produce market buildings (but estimated that they will need no more after 1987); in Hong Kong, old markets are being renovated and new ones are being built, both in commercial and residential areas; Taipei is planning seventy-four new fresh-produce market buildings, and in Harare, twenty-one 'people's markets' are planned and some have been built.

In a number of these cases, however, the aim of the market building programme is linked to that of relocating street traders in order to 'neaten' the city. In very few cases is street-market infrastructure provided on a permanent basis (the Pettah bazaar in Colombo is a notable exception to this) and in most cities street markets are seen as a way of formalizing or containing street traders until such time as formal market-structures can absorb them.

Because of the nature of markets policy in most of the cases considered here, the issues discussed below primarily reflect official attitudes and practices with regard to formal, rather than informal, market infrastructure. A discussion of the central issues follows.

The nature of infrastructural provision

In most cases, new markets are being provided with relatively high standards of infrastructure and services. This is the case particularly in Singapore, Hong Kong, and Taipei. In Singapore, for example, public markets are provided with specialized stalls: meat and fish stalls have cleaning and preparation surfaces and access to refrigerated storage; cooked-food stalls have cooking and washing facilities and communal tables and chairs for customers; and lock-up kiosks of 2·5 by 3 metres are provided for the selling of clothing, jewellery, materials, and so on. Stalls are provided with individually metered water, gas, and electricity. There are communal toilet facilities and cleaning and garbage-removal services. In some markets, stalls even have their own telephones.

It is now official policy in a number of cities to regard the provision of markets as being part of basic urban commercial and residential infrastructure. In Singapore, Hong Kong, and Taipei, the provision of a fresh-produce market is considered an integral part of any new housing development and many office and commercial blocks have the first two or three floors turned over to market use. In Colombo, Bombay, and Harare, while markets are not automatically integrated into new developments, they are nonetheless viewed as an essential public service and are thus accorded some priority in terms of city development.

The provision of infrastructure and servicing in on-street locations occurs at a far lower level. Most authorities provide some basic cleaning and garbage removal services for officially acknowledged street markets, but the provision of shelter and selling surfaces is usually left to individual sellers. This is largely because most of these markets are viewed by the authorities as temporary and therefore they do not qualify for public investment.

The primary impact of this policy is seen in terms of its effect on the cost of market provision – both to the individual stall-holder and to the public authority. These issues will be discussed below in the following two subsections.

The effectiveness of service and infrastructural provision varies considerably between as well as within cities. The older markets in most cities are less well serviced and (except in Hong Kong) there has been little attempt to renovate these markets. In street markets, the extent to which city cleansing and garbage removal are effective is often dependent on factors other than the level of service. In Bombay, for example, many of the pavements are not hardened and this means that waste fruit and vegetable matter is trodden into the soil and is difficult to remove. Where inadequate toilet facilities are available, traders simply use the streets and gutters and this can create major hygiene problems. Where there is overcrowding and congestion of street traders (particularly where the selling space has for some reason been restricted), facilities and services become overloaded and are no longer able to cope.

Financing of public markets

In most of the cases considered here, public fresh produce markets are either viewed as a public service and no attempt is made to generate profits from them or, while the intention is that they should cover costs, it is nonetheless accepted that they must be subsidised.

Thus, both in Colombo and Bangkok, they are viewed as a public service. In Colombo, all markets run at a loss and receive large subsidies. In this city, a standard rent is charged in all markets regardless of their age, location, or the level of service provided in them. Significantly, the new 'supermarket' currently being built in the Pettah area using a government loan will have to charge rents sufficient to cover the capital and interest repayments.

In Singapore and Hong Kong, it is official policy that the revenue from markets should cover costs, but it has never been possible to achieve this. In Singapore, annual budgetary allocations are made for market operation and construction, and these loans are repayable over 60 years at an annual interest rate of 7·75 per cent. This represents a large subsidy to market provision: it is estimated that the (1983) capital cost of providing one market stall was S$40,000 (£10,800), while the annual rental received from the most expensive stalls (the cooked-food stalls) was S$960 (£259). Annual revenue was therefore less than 25 per cent of the annual interest repayments on the capital cost. In the past, this subsidy has been justified on the grounds that it is a form of public-health promotion. However, an escalation of building costs and interest rates has necessitated a reconsideration of the subsidy and rent levels, and is undoubtedly influencing the government's decision not to build more markets after 1987 (see Table 1).

Table 1 Income and expenditure on public markets in Singapore, 1980–2 (£)

	1980–1	*1981–2*
INCOME		
Rentals	2,867,346	3,131,908
Other income	96,685	157,894
	2,964,031	3,289,802
EXPENDITURE		
Salaries, wages	1,171,033	1,433,357
Maintenance	644,383	696,155
Property tax	472,925	484,949
Adjustments	343	–
Loan charges	1,636,506	1,752,532
	3,925,190	4,366,993
DEFICIT	961,159	1,077,191

In Hong Kong, financial allocations to the Urban Services Department (for built markets in the commercial areas) and to the Housing Department (for built markets in residential areas) originate from different sources. The Urban Services Department draws its funds directly from central government. The markets are supposed to be self-financing, but in practice, some 40 per cent of costs are subsidised. The official reason by which this is justified is that these markets contain primarily street vendors who have been relocated. Vendors have to adjust to a situation in which rents are much higher and they are also more likely to face competition from new street sellers who take their place (illegally) outside the market. Markets in the residential areas are financed directly from the housing fund. These markets are, it is claimed, generally more successful financially, because it is easier to prevent competitive informal markets from establishing themselves in the housing estates.

Taipei is the only city of the seven cases considered here where formal markets are financed on a cost-recovery basis. Finance for market construction is provided by the municipality, and in recent years there have been shortages of funds at this level. Stall-holder rentals are designed to cover capital and interest redemption on the building, calculated on the basis of 4 per cent of land costs and 10 per cent of the replacement cost of the building annually, and rents are raised each year to keep pace with escalating replacement values. All other market services operate on the basis of user charges.

In almost all cases, the considerable drain on city finances represented by fully built, high-service markets is leading to a critical review of this form of provision.

Stallholder costs

The level of stall rentals in market buildings is directly related to financing policy and, in particular, the attitude of the authorities towards cost recovery. In Colombo, Bombay, Harare, and Bangkok, stall-holder rents in market buildings are low and relatively uniform. In Singapore, rents are charged according to the type of stall occupied and are based on the anticipated average profits generated in each category of use. Private markets in Singapore allocate stalls on an annual auction basis, and because of the high demand for space in markets less affected by stringent government regulations, auction prices have reached S$9000 (£2400) a stall. In Hong Kong, stall rents in formal markets are made up of two components: a minimum rental which is updated each year and which is supposed to cover the cost of capital and interest redemption and operating costs; and a 'premium' factor which is the amount above the basic rent which a stall-holder is prepared to pay. This premium is determined by a process of tendering for stalls, which occurs every three years. The result is that in housing estates particularly, rents for stalls were very high and reached HK$3000 (£272) a month. In Taipei, the principle of cost recovery is more strictly adhered to, and as a result stall rents are also high. Thus, for example, the occupier of a small (1·5 square metres) fruit-and-vegetables stall could pay NT$700 (£9·58) a month, and in addition, paid user charges for water, garbage removal, and electricity. Finally, in Harare, the main rationale underpinning stall rentals is simply to keep these as low as possible: the charges are nominal only. However, a theoretical distinction is made on the basis of location and degree of cover: the charge for an open site in a people's market in 1983 was the equivalent of 6 pence a day; for a covered site, 11 pence; and 22 pence for a site in the new market located on the edge of the CBD.

In all cases, rents paid for street locations are lower than for built markets and are usually covered by the licence or registration fee. In some cities (such as Hong Kong) an additional 'fixed pitch' fee is levied on permanent street hawkers.

Implications of market finance policy

The basic aim of the markets policy in each of the cases considered – that market provision should be linked to a relocation of street hawkers, together with the practice of providing high levels of infrastructure and servicing in a built market

– has had a major impact on informal traders. Inevitably, it has only been the larger, more established, and permanent street traders who have been able to make the transition to a market building location with its higher rental and other financial commitments. The smaller, intermittent street trader has frequently been eliminated in the relocation process, or further marginalized as a result of the removal of designated street-trading areas and established street markets. Moreover, in order that traders can operate successfully in the built-market situation, it becomes necessary for the authorities to attempt to eliminate the competition provided by other street traders operating in cheaper street locations outside the market building. Smaller traders are thus subjected to a greater degree of harassment as a result of the markets policy. Clearly, such a policy is incompatible with one which aims to alleviate unemployment and poverty through the promotion of informal trading.

Further, it has proved almost impossible in most cities considered here to provide markets on a cost-recovery basis, as even more established street traders are often unable to afford the rents which would be required to finance the sophisticated and elaborate infrastructure in market buildings. The authorities are therefore forced into situations in which subsidies have to be provided if the market is to function at all. Alternatively, as is the case in Taipei, every effort is made by hawkers to return to or remain in a street location rather than relocating to a market building.

In most cases, the lower level of infrastructure provided in street-market locations is far more appropriate for the needs of informal traders. There is also a greater possibility in these markets of making informal service arrangements (for example, having water and electricity from neighbouring shops), and there is greater flexibility in terms of the use made of urban infrastructure by informal traders.

Stall-allocation methods

Information on stall-allocation methods in the seven case-study areas was not always available for all types of markets, and was particularly difficult to obtain for less formal and for informal markets. Generally, however, operators within wholesale markets exhibit greater continuity and enjoy greater security of tenure than those in the retail sector. In those markets for which information was available, there appear to be three main methods of stall allocation: the auction or tender method, the ballot method, and the first-come, first-served method.

There is a tendency for the auction method to be used where profit margins are higher – for example, in Bombay's main produce wholesale market (Crawford market). Here, places were sold many years ago and prospective new entrants are unable to enter the market. Allocation by auction also occurs in Colombo, where access to the beef wholesale market is determined by auction every 3 years. In Hong Kong's retail and wholesale public markets, all stalls are allocated by auction: stall rent is divided into a 'minimum rent' and a 'premium rent'; with the latter being determined by an auction every 3 years. Inevitably, the effect of the auction method is to push up the cost of stall rentals and to exclude the poorer trader from market locations.

In Bangkok's wholesale markets, space is rented on an annual basis, but policy is to give preference to existing operators. Producers are charged an entry fee and operators a monthly rental and also a premium 'key' payment which is theoretically a contribution to capital costs: the legality of this widespread practice is questionable. A widespread complaint in all these cases is that, while reasonable continuity is desirable, the allocation of sites, effectively in perpetuity, causes an excessive build-up, in the face of no new outlets being created, of goodwill or windfall profits.

In a number of cases, the ballot method is used to allocate stalls. In Colombo, wholesale fruit-and-vegetable stalls were balloted on a one-off basis. Rights now exist in perpetuity and may be passed on to children, provided nominal monthly rental charges are met. Some of the larger retail markets also operate on this basis. In Singapore, public retail markets are divided into use sections, and applications for stalls in each section are allocated by the ballot method. A limit of two adjacent stalls is placed on each family. Similarly, in Taipei, public markets are divided into use sections. Lists of street traders are provided by the police and the ballot system is used to allocate these traders to stalls in their respective use zone. If there are insufficient traders of a particular type to fill a use zone, traders of other goods may be placed there.

While the ballot method is potentially fairer in terms of allowing access to stalls, and does not discriminate against poorer traders, it can give rise to problems, particularly that of undesirable use mix. This is most evident in Taipei markets where it is not uncommon to find a clothes seller allocated a stall in the middle of the fruit-and-vegetables section, or a jewellery seller in the middle of a row of meat sellers. A further problem has to do with the tenure period under the ballot system: where stalls are granted to a lessee for an indefinite period, newcomers to the market are discriminated against unless supply of stall space exceeds demand.

Part of Musika wholesale market in Harare provides an example of the third method of stall allocation: first-come, first-served. The Musika market is divided into three sections: a permanent section, a section where stalls are allocated on a weekly basis, and a section where they are allocated on a daily basis. In the last two sections, the first-come, first-served method is used, but preference is given to those who can show, by producing a receipt, that they occupied a stall the previous week, or day (depending on the section). Production of a receipt can also ensure applicants access to the same stall they previously occupied, provided that they wish this. In the case of the daily market, the daily fee charged was Z$2·50 (£1·14). Most of the operators using this sector are small producers. As pressure on this space has grown, allocation has increasingly occurred politically, through pressure from councillors. Theoretically, the dominant criterion is 'need', but reports of allocation through bribery, or for reasons of political patronage, are widespread.

The Musika system offers a reasonable degree of flexibility: the different leasing periods offered recognize the differentiation which occurs amongst informal traders, both in terms of their degree of permanence, and their ability to pay for selling space. The system also allows for variations in turnover amongst stall-holders, and newcomers can, by arriving at the market early enough, gain access to space.

In most cases, retail markets, and particularly the more informal ones, operate on a first-come, first-served basis, where daily, or sometimes even half-daily, stall tickets are sold. Unquestionably, this system is the most equitable and most satisfies the requirements of the smaller, economically more fragile vendors.

Wholesaling

The food wholesaling system exerts a powerful influence over the nature and pattern of small-scale vending. It is apparent from all the cases in question that the wholesaling system represents a key policy instrument in either promoting or retarding small-scale food distribution. Observations relating to significant aspects of this phenomenon are grouped under the following heads: degree of specialization; centralization; location; linkages; ownership; selling agents; buying agents; entry procedures; issues of layout; and market organization.

Degree of specialization

In all cases, varying degrees of specialization occur. The most common form is a separation between meat, fish, and fruit-and-vegetables wholesaling. Thus, in Harare, all meat sales are controlled from one abattoir and auction yard, fish is imported by a number of independent agents, and fruit and vegetables are distributed from two supply areas. In Bombay, meat, fish, and fruit-and-vegetables wholesaling occur separately, as they do in Colombo, Bangkok, Hong Kong, Singapore, and Taiwan. From the point of view of the consumer, however, the system works best when there is easy access to a full range of products. This is best exemplified in Colombo, Sri Lanka. All three forms of wholesaling occur within a few blocks of each other in the central Pettah area: in effect, they comprise an integrated system offering easy access.

Within the fruit-and-vegetables wholesaling markets, a wide range of products is sold in all cases, again to the benefit of the consumer purchasing for retail purposes, since it is possible to purchase most required products under one roof. Perhaps the broadest range is found in the Mahatma Joytiba Phule (Crawford) market in Bombay: separate sections exist for fruit and vegetables, poultry, spices, wooden articles, curios, household goods, and pets (monkeys, birds, and tropical fish).

In the case of Bangkok, five fruit-and-vegetables wholesale markets exist. Although the basic products sold from each are the same, each has taken on an area of specialization related to the agricultural focus of the sector of the city within which they are located (thus, one specializes in pumpkins, one in oranges, and so on).

Centralization

An important issue relating to the wholesaling system is the degree of centralization which occurs. It is necessary here to distinguish between economic and spatial centralization.

In Harare, two relatively separate fruit-and-vegetables wholesaling systems operate, reflecting the colonial history of the city. One system, which is orientated primarily to the CBD and to the outlying shopping centres serving mainly higher-income white residential areas, is located within the CBD and is economically centralized, in the sense that it is dominated by a very limited number of agents. The other, larger, system is orientated towards the numerically dominant black population and focuses on the Musika market. Here, the system is highly decentralized: it is made up of a great many farmers and agents.

A characteristic of most of the Far Eastern cities is that the food wholesaling system is economically extremely decentralized: the five fruit-and-vegetables wholesale markets of Bangkok, for example, house in the order of 900 operators. There is no doubt that the greater the degree of economic decentralization, the less the control of any one set of agents or operators, with favourable consequences for prices. This is borne out, for example, in Bombay, where the wholesaling system in the Crawford market is controlled by a relatively small number of agents. Complaints that these agents exploit both farmers and customers are widespread.

Spatially, the situation varies. As stated, in Harare almost all fruit-and-vegetables supplies are channelled through two spatially discrete outlets. In Bombay, activity is highly centralized in the downtown Crawford market. In Colombo, too, almost all produce is channelled through the Pettah markets. In Bangkok, five fruit-and-vegetables wholesaling markets exist. However, 500 of the approximately 900 wholesalers operate from the Pak Klong Talad–Yod Piman market complexes (theoretically, two separate markets, but operationally a single complex). In Singapore, 75 per cent of food is imported, and this encourages a degree of centralization: two major fruit-and-vegetables outlets exist on Maxwell and Hong Kong roads. In Hong Kong, the pattern is more decentralized: outlets exist near the central area, north Kowloon, and Kennedy town. In Taiwan, all produce is channelled through a single, centrally located facility.

In most places, a second level of wholesaling, made up of independent operators running wholesaling 'shops', has arisen in response to the limited number of wholesale markets. These operators buy in bulk from the market and then transport produce to different subsectors of the city to supply retailers: again, the system increases costs. In the case of Singapore, Hong Kong, and

Taiwan, open-air wholesale street markets occur at night, between 2000 and 0500 hours: second-level wholesalers purchase from the main markets and then re-sell to retailers.

Excessive spatial centralization gives rise to a number of negative consequences. One is poor traffic circulation. Wholesale markets generate a considerable amount of traffic. Thus, for example, the amount of produce entering and leaving the Pak Klong Talad–Yod Piman complex annually is in the order of 250,000 tons. In this case, some 60 per cent is transported by truck and 40 per cent by water, but in most cities, vehicular portage is overridingly dominant. In highly centralized systems, all movement is focused on a single point and congestion is frequently the result, particularly when that point is centrally located within the city. Crawford market in Bombay is an example of this: market activity jams central-city movement for many blocks, a situation aggravated by the fact that a considerable amount of movement is by animal cart, which is slow and cumbersome. Similarly, the Pak Klong Talad–Yod Piman complex in Bangkok is located near the Memorial Bridge over the Chao Phya River, over which 80 per cent of all traffic crossing the river passes. It has been estimated that only 14 per cent of traffic at peak market time on Chakapet Road is market-related (Kuhn 1972). The resultant congestion significantly impairs the efficiency of the market, to the degree that an alternative, even more centralized substitute, which is considerably less favourably located in terms of consumers, is being considered.

Ameliatory attempts to control the problem of congestion frequently have unexpected and negative consequences. Thus, for example, in Bangkok, trucks transporting food to markets may only operate within city limits between 2100 and 0600 hours. Frequently, however, for a variety of reasons, they are unable to complete unloading by 0600 and are forced to park until 2100 before they can return to the countryside. This is not only inefficient and costly in time and marketing terms, but the trucks take up parking intended for customers loading produce. Parking near the market is extremely difficult to find.

In the case of Colombo, although the market *system* is centralized in the Pettah area, the decentralized nature of facilities within this diffuses traffic throughout the area and reduces congestion. Despite this, plans are afoot to centralize the entire wholesaling system under a single roof. Inevitably, in order to control congestion and to find sufficient space, such a facility must be located well away from the centre of the city, in a location which is accessible only to those who have access to private vehicular transport.

A second consequence of excessive centralization is its impact on small farmers. This is clearly demonstrated by the case of Bombay. Much of the food is produced by small farmers located beyond the periphery of the city. Output per production unit is relatively low, and consequently transport costs assume a relatively greater proportion of total costs than they do in the case of large producers. With the lateral growth of the city, the Crawford wholesale market has become increasingly inaccessible to the small farmers: it is uneconomic for them to transport small quantities frequently over large distances in order to generate small turnovers. In this situation, agents who collect and transport the output of a number of small farmers have emerged. Because of the dependence of small farmers on these agents, they effectively control the production system and exert a powerful influence on prices.

Similarly, centralized systems ultimately disadvantage small vendors, who play a vital role in the food distribution system of Third World cities (for example, in the Pak Klong Talad–Yod Piman complex in Bangkok, 24 per cent of total produce is purchased by small vendors). These vendors have to travel considerable distances to the market in order to purchase supplies to take back to the more peripheral areas they serve, which are frequently far closer to actual points of production than the wholesale market. This forces up costs and relatively disadvantages very small vendors in relation to those who have access to private transportation. In Bombay, attempts have been made to alleviate the problem somewhat, by reserving the last carriage of passenger trains for vendors' wares, but the wholesale market is not directly related to the fixed rail-line system.

This problem of transhipment can reach severe proportions. In Thailand, Bangkok operates as the food transhipment point for most of the country: provincial wholesaling outlets do not exist. Some 35 per cent of vegetables and 46 per cent of fruit sold from the Pak Klong Talad–Yod Piman complex is re-shipped to the provinces where most of the growing occurs, and 90 per cent of transportation occurs by truck (Kuhn 1972).

Finally, the phenomena of economic and spatial centralization are frequently interrelated. One way in which spatial centralization promotes economic centralization of the produce-supply sector has already been outlined. Another, which relates to the centralization of the distribution system, results from a shortage of space within the wholesale outlets themselves. Inevitably, in centralized locations, space is scarce, as, for example, is demonstrated in the Crawford market in Bombay. A limited number of agents can operate from the market and its constantly increasing range (fuelled by a lack of alternative outlets) places the traders operating from it in an increasingly oligopolistic position, with negative consequences for prices. A similar situation developed in the formal Musika wholesale market. As demand grew, however, an informal wholesale market, consisting primarily of farmers, grew up on vacant land on the other side of the retail market, and the pressure has been relieved somewhat.

Location

Ideally, in their location, wholesale markets should reconcile the interests of two primary sets of users: food producers and consumers. All the cases reflect, to a greater or lesser degree, the consequences which result when the expansion of the wholesaling system is not carefully co-ordinated and integrated with city growth. Initially, in all cases, wholesale markets were centrally located and were easily accessible to both producers and consumers. With increasing urban growth, the degree of accessibility to the majority of both users and producers has declined. Further, the pressure on the centralized facility has increased, with all the consequences outlined above. In most cases, when the pressure has become really severe, relocation is proposed. Frequently, the relocated facility is seen as a total *substitute* for the old market: the market is moved in its entirety, away from the most central location. Because of this, accessibility is increasingly defined, not in absolute terms but in terms of ease of vehicular access, and this detrimentally affects poorer users. Thus, an eccentrically located market facility is being planned in Colombo, to house the entire wholesaling system of the central Pettah

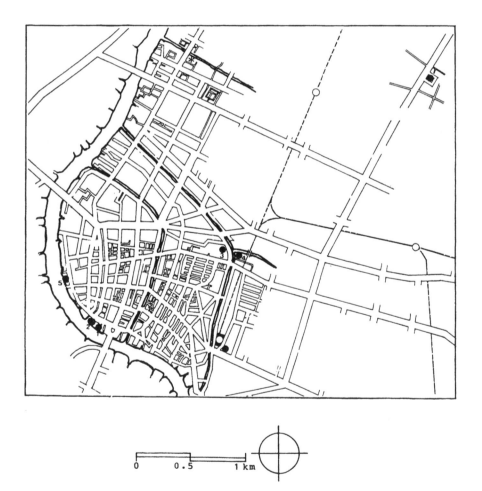

Figure 25 Bangkok wholesale markets: 1. Yod Piman; 2. Pak Klong Talad; 3. Mahanak; 4. Padung Krung Kasem; 5. Public Warehousing Organization

area. In Singapore, a structure to centralize all fruit-and-vegetables wholesaling is being built on Carpenter Street. Similarly, even though the pattern of wholesaling in Bangkok (where the five wholesaling markets serve a large area of the city and are all located on subregional roads and waterways – Figure 25) is spatially the most decentralized, proposals have been made to centralize the entire wholesaling system into one complex (Kuhn 1972). The case of the Pak Klong Talad–Yod Piman system also illustrates, as has been discussed, the need to desegregate wholesale facilities from other large traffic generators.

Clearly, developing a phased system of decentralization, where additional markets (as opposed to substitutes) are carefully integrated with urban growth and where a constant relationship between agricultural and urban areas is maintained, is an important part of urban management.

Linkages

All the cases observed reflect the fact that wholesale markets are important urban generators and inevitably spawn a variety of services around them. The most common of these are restaurants and food take-aways, banks and land banks, chemists, barbers, and other forms of personal service, services related to agricultural activity (for example, the selling of seed, fertilizer, and insecticides) equipment for handling and transporting produce, baskets and packaging materials, and so on.

An almost ubiquitous association, too, is that between wholesale and retail markets. The retail markets attach themselves to the wholesale market to break bulk. These markets sell directly to the consuming public but primarily serve small vendors who purchase relatively small quantities and who form an important part of the food distribution network in all Third World cities. In Harare, for example, the Musika retail market, which lies between the formal and informal wholesale markets, houses some 2000 operators over weekends. In some cases, time zoning has been introduced to prevent the operation of the retail market from interfering with unloading at the wholesale market. Thus, the retail fish market attached to the fish wholesaling outlet, as well as the fruit-and-vegetables retail market in Pettah, Colombo, can only operate after 0900 hours.

Ownership

The ownership of the wholesale markets varies between public and private. In Harare, the outlet on the fringes of the CBD is private and the Musika market is municipally owned. Similarly, in Singapore, one fruit-and-vegetables market is private and one is public. In Bangkok, three of the five fruit-and-vegetables wholesalers are privately owned and the other two municipally-owned. In Colombo, Hong Kong, and Taiwan, markets are almost entirely municipally run.

Selling agents

In most of the wholesaling systems, both agents and growers participate. Thus, in the case of Bangkok, which is the only city for which statistics are available, 54 per cent of vegetables and 45 per cent of fruit are sold directly by growers, while 38 per cent of vegetables and 36 per cent of fruit are sold by assembly traders from the villages. Shipments from commissioned agents make up the rest (Kuhn 1972). In the Pak Klong Talad–Yod Piman complex, however, there is no special farmers' market. Farmers either have to struggle for space within the market by subleasing from other operators, or more commonly, they sell from the parking area – a pattern which substantially increases congestion. In the case of Bombay, the wholesaling system is controlled entirely by agents. The degree of agent participation in all cases is directly related to the degree of difficulty experienced in transporting goods to market.

Buying agents

In all cases, smaller wholesalers, formal retailers, informal vendors, and the consuming public purchase from the main wholesale markets. Direct-consumption purchasing comprises by far the smallest proportion. Significantly, sales to vendors in all cases form an important part of market activity. Thus, in Bangkok, 48 per cent of vegetables and 39 per cent of fruit are purchased by smaller wholesalers and formal retailers, while 24 per cent of total produce is sold to vendors (ibid.).

Market entry and stall-allocation procedures

This is discussed in the section outlining the management and administration of markets.

Issues of layout

The layout of, and infrastructure provided in, wholesale markets, can obviously take many forms, and decisions relating to this are governed primarily by climate and economics. Of critical importance in determining efficiency of operation, however, is the vehicular circulation system. Because of the volumes of produce handled, the loading and unloading of vehicles is of considerable importance. The entire Pak Klong Talad–Yod Piman market complex, for example, only has nine access and egress points and only two of these can be penetrated by vehicles. The bottlenecks which result are enormous and contribute significantly to marketing costs.

Market organization

In most cases, markets are controlled by a market master and an administrative staff complement. The only exception to this, and one which operates extremely well, is the Musika market in Harare, Zimbabwe. Although the market is municipally owned, operationally the market is entirely internally run by a system of committees: all decisions, from fees and entrance charges, to garbage removal, to policing, are made by people operating from the market, in consultation with all affected interests.

Chapter 4

Market cases

This final chapter examines some of the markets analysed which provide the basis for the discussion in Chapter 2. Markets have been chosen to illustrate the issues identified in Chapter 2, and, in each case, examples of markets which operate poorly and markets which operate relatively well are shown.

Market size and location

The size and the location of markets are inextricably connected and affect market performance significantly. Certain markets are too small to provide a sufficient 'magnet' to potential customers. Telford Gardens market and Ho Man Tin market experience this.

Markets which are too small

Telford Gardens market, Hong Kong

This is a privately developed market building located in a middle-income residential area. It is part of a larger flat and shopping-arcade complex, and it retails fresh fruit and vegetables, fish, meat, chicken, and general consumer goods. The market, which is a square in form, has an area of approximately 480 square metres and contains forty-two stalls.

Overall, the market is too small. Given the fact that it is somewhat removed from major movement channels, it needs to attract custom in its own right and its limited size mitigates against this. This is exacerbated by the fact that the range of goods offered is wide: the number of stalls in each use category is thus very few. In these situations, a vicious cycle is set in motion. The lack of patronage causes vacancies (several stalls were vacant at the time of observation), which reduces confidence and thus patronage still further. Moreover, since the market is privately owned, it operates on a profit basis: capital and redemption costs, operating costs, and profit need to be recovered from a declining rent base. This continually forces up unit rentals which, initially, escalates product prices and, ultimately, hastens the advent of economic mortality.

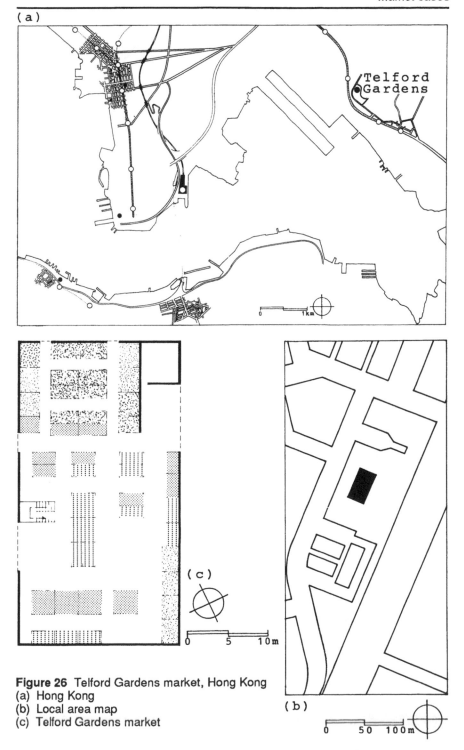

Figure 26 Telford Gardens market, Hong Kong
(a) Hong Kong
(b) Local area map
(c) Telford Gardens market

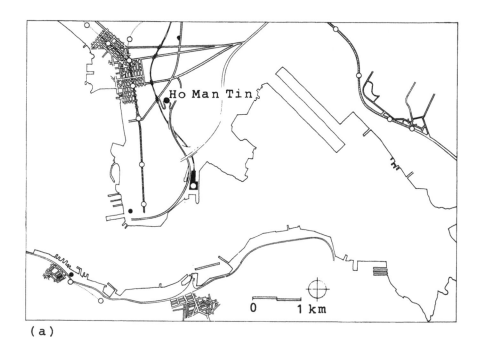

(a)

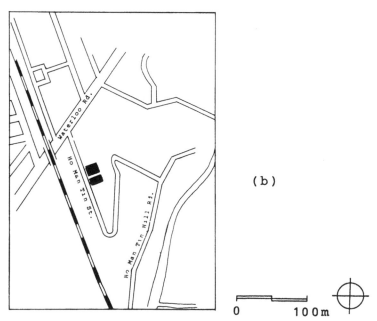

(b)

Figure 27 Ho Man Tin market, Hong Kong
(a) Hong Kong
(b) Local area map

(c)

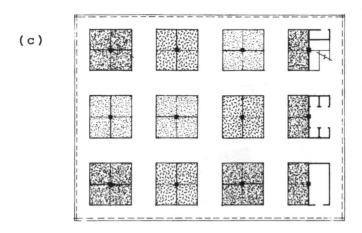

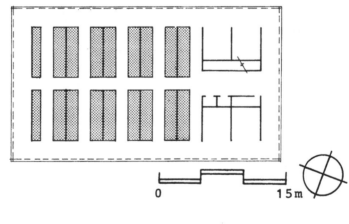

0 15m

(d)

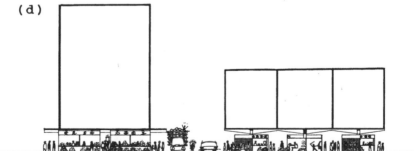

(c) Ho Man Tin market
(d) Ho Man Tin market cross-section

115

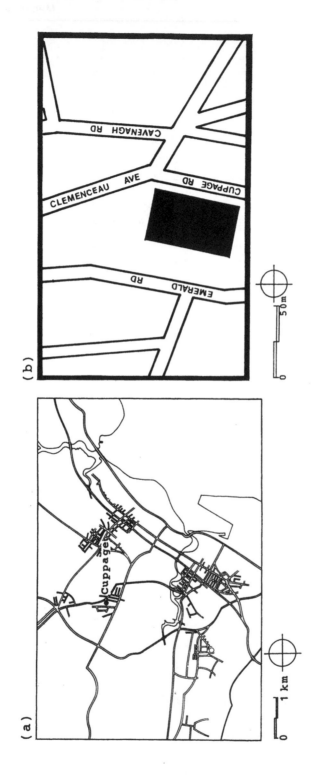

Figure 28 Cuppage market, Singapore

(a) Singapore
(b) Local area map

(c)

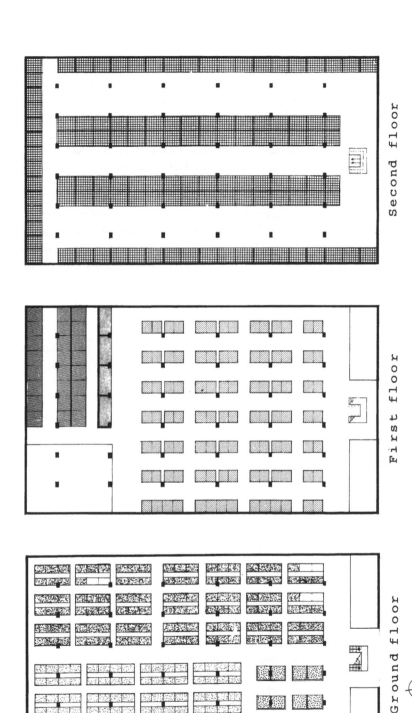

Ground floor   First floor   Second floor

0   10m

(c) Cuppage market

Ho Man Tin market, Hong Kong

This is a retail fresh-product market containing fruit, vegetables, fish, meat, and poultry sellers. It is located within a housing estate on the ground floors of two adjacent apartment blocks. The market is therefore divided into two separate sections: a 'dry' market selling fruit and vegetables and a 'wet' market selling meat and fish. The dry market is 788 square metres in size and contains thirty-six stalls. The wet market is 585 square metres in size and contains forty stalls.

Its locational inward orientation to the housing estate (as opposed to an external orientation to routes connecting a number of such areas) ties its fate to the fortunes of a very local market and increases its economic vulnerability. Additionally, and in part resulting from this, it is too small to reflect a significant generative or attractive power. In this case, the fact that the market specializes in a limited number of products works to its advantage.

Markets which overcome size problems

Markets can operate successfully at a wider range of sizes, provided their location and design is appropriate. Cuppage market and the Yaowarat Street market system are examples of successfully performing large markets, while Dadar and the Kollonawa Road markets are much smaller in extent.

Cuppage market, Singapore

Cuppage Road market is strategically located at the edge of the central-business area of Singapore. Because of its centrality it is in great demand. It has three floors which are each 2100 square metres in extent, with a gross floor area of 6300 square metres. The market concentrates almost entirely on the sale of fresh produce and food. The ground floor contains 200 meat, fish, and poultry stalls; the first floor 100 fresh fruit-and-vegetables stalls, twenty-four clothing kiosks, and four frozen food stalls; the second floor is made up entirely of cooked food kiosks – there are 116 of these, together with fixed tables and chairs. The market thus contains 444 stalls in all. This amount of activity, initially made possible by the central location, has in turn attracted custom over an increasing and now large trade area. Vacancies are minimal. The difficulty of vertical separation has been overcome by the extent of specialization and the scale of each activity, which enables highly selective buying. The cooked-food centre, which is not dependent upon impulse buying, draws customers upward through the market. The most vulnerable section of the market is the clothing section where the mass achieved is not of the same scale as the other products.

The Yaowarat Street market system, Bangkok

The Yaowarat commercial area of Bangkok represents an example of a concentrated street-market system and illustrates a number of important issues

Figure 29 The Yaowarat Street market system, Bangkok
(a) Bangkok
(b) Local area map
(c) Yaowarat Street market system cross-section

and relationships: the relationship between the intensity of movement flow and selling function; the relationship of different elements within the retailing system and particularly that between wholesaling and retailing; the importance of specialization; and the creative use of existing infrastructure.

The largely continuous market system extends some 1·25 km across four large city blocks and is 0·7km in width at its widest point. In places, it extends vertically up one floor as well. It occupies one of the central commercial areas of the city and the intensity of movement and activity are considerable. The system contains three main elements: small formal shops, wholesale sheds, and street sellers. Spatially, major city thoroughfares, such as Chakraphet Road, Chakra Wat Road, and Ratchaway Road, are free of street-market activity and thus create breaks in the system: the many smaller streets and passages internal to these blocks, however, accommodate street sellers as well as vehicles and other forms of transportation.

The system reflects a high degree of use specialization (see Figure 29(b)). In block A there is primarily retail selling of cloth, clothing, jewellery, and accessories with cloth wholesaling occurring from a large covered shed. Block B has a greater concentration of cloth and clothing wholesalers, and on its southern edge, along the klong, there are concentrations of smaller fresh and cooked food stalls, together with tables and chairs. Block C consists primarily of retailers of shoes and household goods and blocks E and F house mainly food wholesalers and retailers, including fresh-produce markets.

There are a number of definable features contained within this general pattern. First, within the general system, high-value goods (for example, clothing and jewellery) occupy the most central and intense locations; lower-value goods, such as fresh produce, gravitate towards the edges. Significantly, the distribution of fresh produce is also affected by movement technology: its gravitation towards the klong reflects the fact that water is an important mode of transporting foodstuffs.

Within individual blocks, identifiable responses to movement flows are discernible. Large bulk activities, such as wholesaling, gravitate towards the centre of the block, with smaller-scale retailing concerns taking up positions directly contiguous to pedestrian flows. Significantly, cooked food and refreshment vendors concentrate at the knuckles where minor internal streets intersect with the major city thoroughfares. The pattern of intensity of activity also varies with locations. In areas of greatest movement flow, vending occurs at the edges and down the centres of streets; towards the edges of the area, it occurs on the two sides of the street only.

While there is a pronounced general pattern of specialization, subtle variations occur within these. In particular, runs of sellers selling similar items, and clearly benefiting from cumulative attraction, tend to break up around a distance of some 100 metres. Frequently, they are replaced by compatible but different uses (for example, clothes sellers are replaced by vendors of shoes or accessories), only to re-emerge some distance on.

An important feature of the system is the mutually beneficial relationship between retailing and wholesalers. Retailers benefit from the accumulation of customers and the ability to replenish stock quickly. Wholesalers benefit, since the relationship helps establish identifiable 'sectors' or districts which assist product search by retailers on a city-wide basis. The consumer benefits from high levels of convenience: search patterns for particular products are clear, while it is still possible to obtain a wide variety of goods over a relatively small area.

The system is primarily pedestrian-orientated. Interestingly, there has been a technological response to the intensity of movement. Selling outlets are serviced primarily by scooters with trailers which move slowly through the pedestrian flow. This is markedly different from many other cases where a concern with accommodating higher technological forms dramatically reduces the intensity and thus the commercial success of the market system.

Finally, the case reflects a sensible low-cost approach to the provision of shelter by making maximum use of existing infrastructure: canvas and forms of coloured cloth cover are strung intermittently between buildings over internal streets to provide shade and cover (the varying degrees of porosity also create varying conditions of soft light); where passages are extremely narrow, perspex and other forms of solid, light-porous cover have been erected; and so on.

Dadar market, Bombay

The fresh produce market at Dadar (Kraneisinhanana Patil market) is relatively small: it measures 48 by 48 metres, giving a total area of 2304 square metres. It is intensively used, however, both by sellers and by purchasers. There are several reasons for this.

First, it is well located. Although it lies outside the main commercial area of Bombay, it is situated in a mixed-use area which is characterized by high-density residential and commercial activities. There is therefore a concentration of potential customers in the vicinity of the market at all times. Further, it is located close to a railway station. This is an important benefit for the fresh produce sellers, as many bring in their goods by rail.

Second, the market specializes in the sale of one type of good: fresh fruit and vegetables. It therefore contains a large number of sellers of a similar type and from the perspective of the consumer, choice, price competition, and the ability to shop comparatively are all maximized.

Third, the form and layout of this market are such that customer flows are concentrated and channelled to all parts of the market: there are therefore no 'dead' or unused areas. Selling is related directly to movement flows and in effect, the market consists of a system of two-sided linear selling elements. Entrances are centrally located on each side of the square opposite the bisecting passages, thus ensuring an even spread of potential customers through the market. While the extension of selling elements which are central to the passages ensures continuity, the central intersecting space is generous enough to allow for easy directional change from one element to another. Additionally, the four open

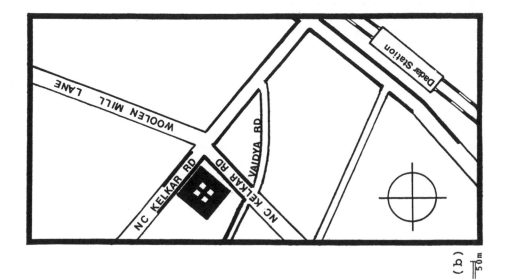

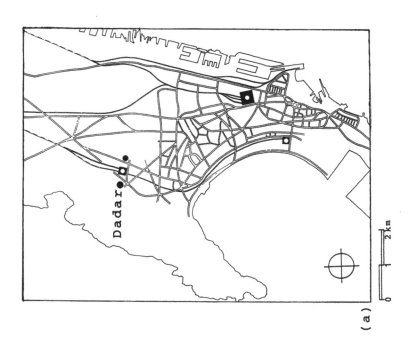

Figure 30 Dadar market, Bombay
(a) Bombay
(b) Local area map

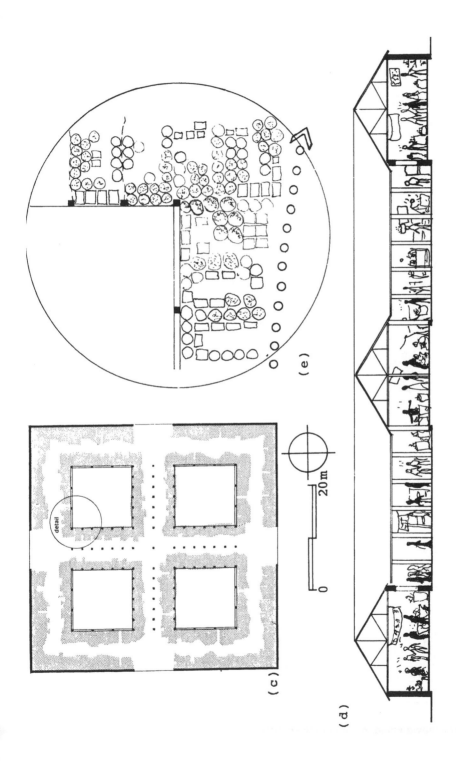

(c) Dadar market
(d) Dadar market cross-section
(e) Dadar market stall detail

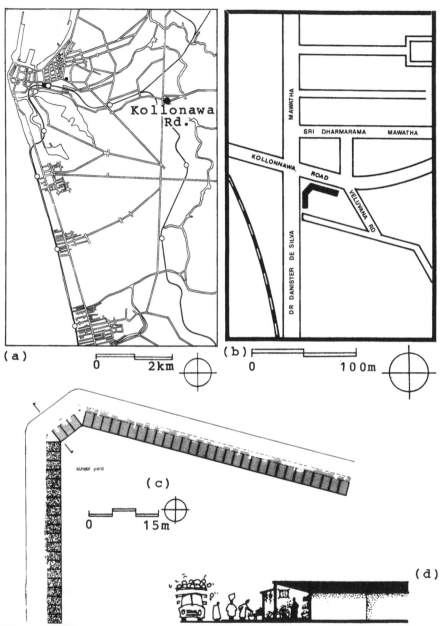

Figure 31 Kollonawa Road market, Colombo
(a) Colombo
(b) Local area map
(c) Kollonawa Road market
(d) Kollonawa Road market cross-section

spaces play an important function. They are unroofed and thus promote light and air penetration and circulation. In effect, since they lie to the back of the sellers, they operate mainly as semi-private spaces: they serve as general storage space, and collective infrastructure, such as communal water points, is located within them. However, public penetration into them is possible with some effort and occasionally they serve as resting places where people meet and talk.

There are no defined selling spaces or publicly provided selling surfaces: sellers provide their own, usually in the form of boxes or sacks. Because of the generally desirable conditions within the market, competition for space is intense, utilized selling areas are small, and the market accommodates a great many traders.

Kollonawa Road market, Colombo

This market consists of a line of municipally built lock-up cubicles. There are fifty-five stalls, selling fruit and vegetables, meat, and some durable goods. In size terms, it is a relatively small market, but its structure and location ensure that it is intensively used. There are several reasons for this.

First, the market is situated at the intersection of two major through-roads which traverse the metropolitan area. The surrounding area contains a mixture of commercial, residential, and institutional land-use, and the market can therefore take advantage of high volumes of both pedestrian and vehicular traffic. Second, it is easily accessible to pedestrian and vehicle traffic. It is possible for cars to pull on to the side of the road and stop directly outside the market stalls, and the relatively wide pavement allows pedestrians either to by-pass the stalls or to browse. Third, the form of the market ensures that it capitalizes on two-directional flows and all stalls are located as directly as possible to these: it treats each trader equitably. Similarly, the L-shaped form promotes natural areas of specialization. In this case, meat sellers concentrate along the shorter arm.

Environmentally, the system operates as an 'urban wall' which reduces the dead space of excessively wide pavements and gives definition to street space. On the internal side, the wall defines a school site. This relationship between market and school, too, is a desirable one: the market contributes towards integrating the school into urban life and promotes informal learning, while the activity generated by the school benefits market activity.

The 'walled kiosk' system obviously caters for somewhat larger permanent traders, but facilities of this kind represent an important element in providing a range of trading opportunities. They are used to great effect in many Far Eastern cities.

Use mix

Use mix poorly handled

In many of the markets analysed the issue of use mix had been poorly handled. Bowrington market provides an example of this problem.

Bowrington market, Hong Kong

Bowrington market is a large, recently constructed market on Wanchai. It is a double-storey building which is divided into two by a roadway. It therefore has four sections: two on one side of the road and two on the other. The market covers 4000 square metres and contains 315 stalls.

While there does seem to have been some attempt to separate out certain uses in the market, in fact there is a great deal of mixing of different and sometimes incompatible uses. Section A sells meat, and fruit and vegetables; section B sells fruit and vegetables, flowers, clothes, and household goods; section C sells fish, and fruit and vegetables; and section D sells poultry, and fruit and vegetables. Section B is highly mixed, with different goods scattered across the entire floor.

There are two main reasons why this has come about. First, it is the policy of the local authority that any newly built market must contain a certain proportion of 'mini' fruit and vegetables stalls to accommodate the smaller hawkers who are being moved from street locations. Each section of Bowrington market therefore contains a perimeter ring of these mini stalls for fruit and vegetable selling, as well as the more specialized, larger stalls designed for other uses. In addition, stalls in section B have been deliberately designed to be non-specialized; all uses other than the meat, fish, and poultry-selling (which are specifically catered for) are therefore jumbled in this section.

A second reason for the poor mixing of uses is to be found in the stall-allocation system. Stalls other than the meat, fish, and poultry stalls are allocated on a first-come, first-served basis and there is therefore no opportunity for specialized areas to emerge. In fact, many stall-holders have apparently applied for two stalls, a specialized one and a 'mini' stall: they then use the mini stall as storage space. Inevitably, therefore, the perimeter rows of mini stalls are 'dead' areas, usually piled with boxes and baskets.

The mixing of different uses in Bowrington market gives rise to a number of problems. First, environmentally the sale of certain goods is incompatible with the sale of other goods, and where conflicts occur there is a negative impact on the entire selling environment. This problem is most evident in section B, where fruit-and-vegetables or flower selling takes place adjacent to clothes and household-goods selling. The sale of fruit and vegetables involves the use of water and generates waste products, which can directly or indirectly affect clothes selling. Similarly, in the other section, meat, fish, and poultry selling generate smells and liquids which detract from a pleasant fruit-and-vegetables trading environment. Chicken slaughtering and selling, which generate

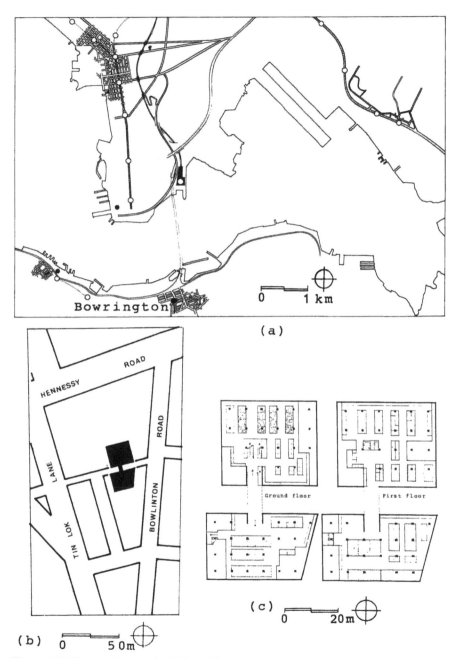

Figure 32 Bowrington market, Hong Kong
(a) Hong Kong
(b) Local area map
(c) Bowrington market

127

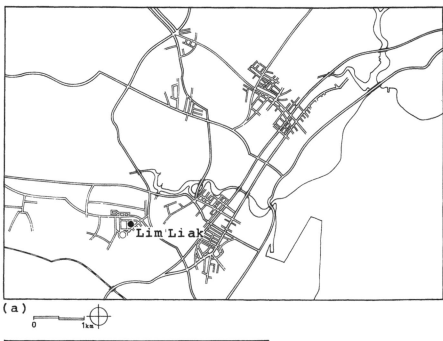

(a)

0   1km

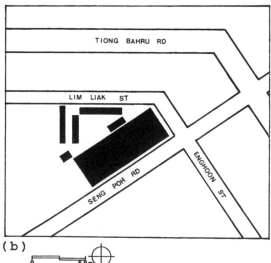

(b)

0   50m

Figure 33 Lim Liak market, Singapore
(a) Singapore
(b) Local area map

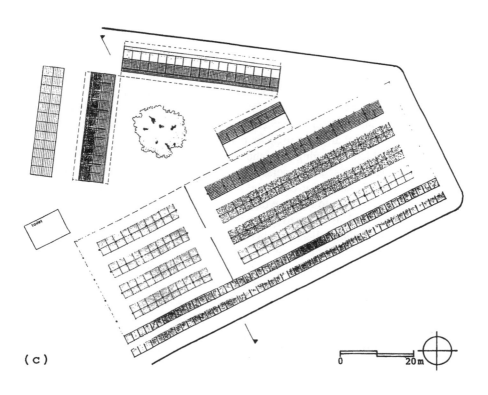

(c)

0   20 m

(d)

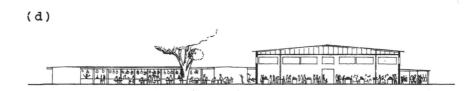

(c) Lim Liak market
(d) Lim Liak market cross-section

obnoxious smells in the slaughtering and de-feathering process are, particularly, incompatible with most other uses.

Second, the sale of different uses requires different kinds of environments and sometimes different kinds of selling infrastructure. This requirement has been catered for to some extent in Bowrington in the provision of specialized meat stalls (with chopping board surfaces and hanging hooks), specialized fish stalls (with cleanable and easily drained selling surfaces and a good water supply), and specialized poultry stalls (with wire cages and chopping boards). However, other goods sold in the market, which also require a degree of specialized infrastructure provision, do not have it. This is particularly the case for the display of clothing, which demands hanging space for storage and display purposes and which benefits from a sense of enclosure. Household goods require plentiful display and storage space, and fruit and vegetables require display space and water supplies. The requirements of all these uses are not met in this case.

A third problem emerging from the system of use mix in Bowrington market is that different uses are dispersed in the market. This makes comparative shopping very difficult: for example, a customer would have to visit all four sections of the market to find the cheapest fruit-and-vegetables stall. Similarly, the customer would also have to search through all of section B to compare prices and quality of clothes sellers. Furthermore, by concentrating together, different uses are able to signal their presence to a potential customer: in Bowrington it is not clear to customers where, for example, household goods or clothing can be found. Finally, by concentrating together, uses can *create* their own environments which are different from the environments of other uses: for example, several rows of clothing stalls together can, by hanging clothes and draping cloth and canvas, create the sense of intimacy and enclosure which is conducive to the sale of these goods. Scattered clothing stalls on their own cannot do this.

Use-mix more successfully handled

There are many examples of both built and open-air markets where the use mix issue has been more appropriately handled.

Lim Liak market, Singapore

Lim Liak is primarily a formal fresh produce market but it also has sections which sell cooked food, clothing, and general household goods. While the issue of use mix is not correctly handled in *all* respects, it achieves a solution to this problem to a greater degree than the market previously described.

The market consists of five buildings and a toilet block. The main building is a large, open-sided market shed (100 by 36 metres in size). The stalls in this section are relatively undifferentiated in terms of infrastructure and consist simply of basic concrete plinths with wooden surfaces. The selling of meat, fruit and vegetables and flowers, and cooked food occurs in this shed with each use occupying broadly separate areas. A 6-metre section on the outer edge of this shed is used for clothes selling. Two rows of wooden kiosks face each other across a 1·5 metre circulation passage, with the high backs of the inner row of kiosks fo ming a dividing wall between this section and the meat- and fruit-and-vegetables selling section behind them.

The north-western edge of the market contains four rectangular sheds housing rows of lock-up kiosks which back on to each other. The inner row of kiosks in three of the sheds faces on to an open triangular courtyard, and the kiosks and the courtyard serve a cooked food/restaurant function. Tables and chairs are placed in the courtyard and the kiosks have been equipped as small kitchens. A wide range of foods is produced by these small vendors.

The outer row of kiosks in two of the sheds, together with the shed in the north-west corner of the market, concentrate on the selling of household goods. In one case, selling is orientated directly on to the road skirting the market (thus avoiding the often-found 'dead edge' situation in formal markets). In the other case, selling is orientated inwards on to a common passageway: canvas overhangs provide shade for much of this passage and allow goods to be displayed in the passage in front of the kiosks.

The section of the shed facing on to the main market hall contains the chicken killing and de-feathering function and it lies in direct proximity to the eating sections (both outside and inside the main hall): the relationship is highly negative, particularly to the cooked food function.

The relative success of this market lies in the fact that different uses are concentrated and are allowed to form different kinds of shopping environments, while remaining integrated into a single market system. Where this separation breaks down (particularly in the case of the chicken-killing section but also, to a certain extent, at the poorly defined and resolved interface between the meat section and the cooked food, and fruit and vegetables in the main hall), the shopping environment operates poorly, and stalls suffer economically as a result.

The different sections of the market are also suited, in terms of their location and layout, to the type of goods being sold. The linear clothing section on the edge of the main hall allows an enclosed environment to develop, completely separate from the fresh produce behind it. The wooden kiosks provide storage space and hanging space, and comparative shopping can take place on either side of a single passageway. The scale of the circulation space and kiosks is appropriate to the kind of environment required for this use.

The outside cooked-food section also works well in that a complete and separate environment is formed. The courtyard is dominated by a large tree which gives shade and scale to the area. Customers seated in the courtyard have visual contact with food sellers on all sides of the triangle and comparative buying is thereby facilitated.

The household-goods section has also developed a distinctive environment. The use of linear infrastructural elements responds directly to pedestrian flows and allows the creation, through canvas overhangs, of a sheltered and protected environment separate from other uses in the market. The lock-up kiosks provide adequate and safe storage space and there is space in front of these to arrange displays.

Pagoda Street market system, Singapore

This is part of an extensive street-market system located in old Chinatown – an inner-city, mixed residential and commercial area. The section depicted shows street selling occurring along a 717-metre stretch of roadway (often on both sides of it), together with a more formal market-building. The street market has developed spontaneously and all the stall infrastructure is provided by the stall-owners. Behind the street stalls are rows of small, formal shops, separated from the street stalls by a pavement.

It is significant that where street traders have been allowed to determine the degree to which different uses are separated or mixed, they have, on the whole, separated themselves into different use-zones. Thus, distinct areas have arisen: a clothing area, a fruit-and-vegetables area, a cooked-foods area, a more mixed zone of fruit and vegetables, spices, confectionery, and household goods, and, in more formal structures, fish, meat, and poultry. Those uses which demand a distinctly different environment – clothing and cooked food, and meat, fish, and poultry selling – all occupy an entirely separate section of the street. Those uses which have only a degree of incompatibility (fruit and vegetables, flowers, spices, household goods, and so on) tend to mix to a greater extent. The formal shops behind the street stalls tend to follow the same usage as the street sellers, although this occurs in a complementary rather than a competitive fashion: the formal shops either provide a wholesaling function for the retailers in the street, or alternatively they may sell a higher quality form of the goods sold in the street.

This arrangement has a number of advantages. First, similar goods clustering together allow the emergence of a total and distinct environment suitable to the sale of the particular goods. From the perspective of the customer, it allows convenient comparative shopping and (by massing) clearly signals to the customer where particular goods can be found. Second, a linear arrangement of stalls (along the sides and centre of a street) minimizes the contact zone between relatively incompatible uses. Third, although uses are separated out, none has been spatially marginalized. The linear arrangement which characterizes open-ended street markets gives each stall equal access to passing pedestrian flows and avoids the danger of 'dead spots' so often found in built markets. Significantly, however, not all streets are subject to pedestrian flows of the same intensity, and uses have distributed themselves in response to these differing intensities. Thus, the cooked-food/restaurant function, which is space-extensive and demands a quieter environment, occurs on a less used side street; higher-value and space-intensive products (flowers, condiments, household goods, and fruit and vegetables) are clustered around the stretch of street where pedestrian flows are most intense – that is, above and below the bend in the road. Clothing, which is more space-extensive, is found at the quieter northern end of the main road.

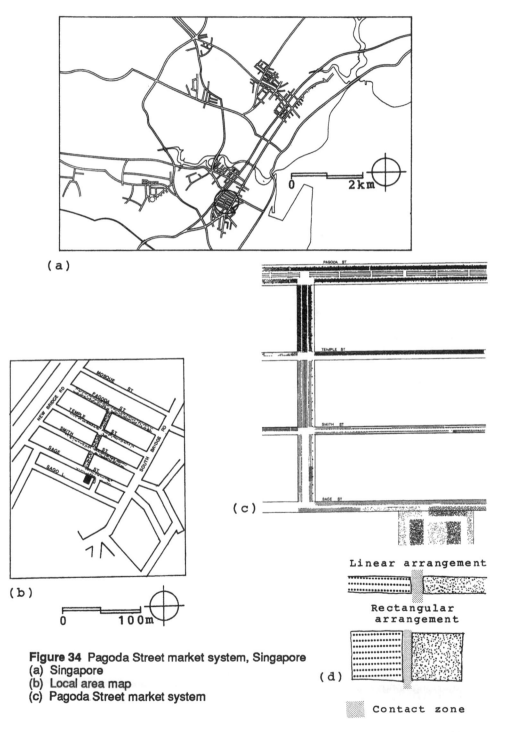

Figure 34 Pagoda Street market system, Singapore
(a) Singapore
(b) Local area map
(c) Pagoda Street market system

133

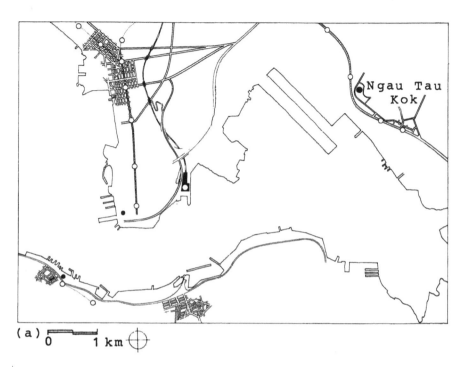

(a) 0 1 km

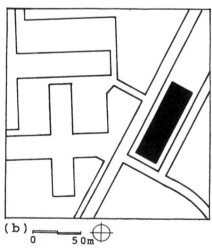

(b) 0 50m

Figure 35 Ngau Tau Kok market, Hong Kong
(a) Hong Kong
(b) Local area map

134

(c)

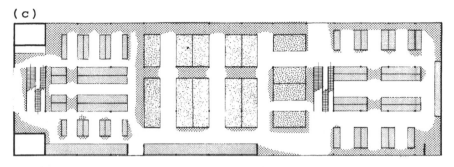

Ground floor

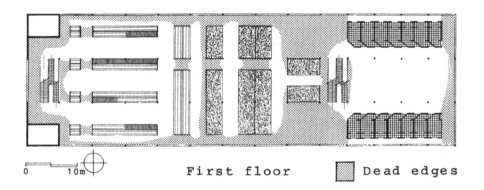

First floor   Dead edges

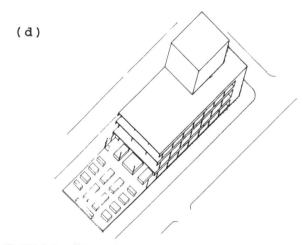

(d)

(c) Ngau Tau Kok market
(d) Ngau Tau Kok market axonometric

135

Market layout

Less successful market layouts

The following two markets provide examples of market layouts with negative characteristics.

Ngau Tau Kok market, Hong Kong

Ngau Tau Kok is a large, retail, fresh-produce market, occupying the lower two floors of a three-storey building. It sells fruit, vegetables, fish, meat, poultry, some groceries, some clothing, and cooked food. The open-sided concrete building provides high levels of infrastructure: stalls are highly specialized, in the sense that they cater for specific uses, all walls and floors are tiled, and there are escalators between the two floors.

This market demonstrates most of the layout problems which are frequently found in formal markets.

Entrance location There are three entrances to the ground floor: one at the south-west end of the building and two opposite each other two-thirds of the way along the building's length. All entrances coincide with escalators to the first floor. A major problem which this causes is that the section to the north-east of the two interfacing entrances is effectively a cul-de-sac: market intensity falls off dramatically here. This is less of a problem on the first floor as the equivalent area houses cooked food operators and collectively these operators draw custom in their own right.

Stall orientation This issue is also affected by the location of entrances. Customers generally seek to minimize effort by entering a market at a point and moving with minimum deviation past a critical mass of stalls (sufficient to allow adequate selection) to a point of exit. In this case, this tendency is frustrated on both market floors by the fact that rows of stalls are, in many cases, orientated at right angles to the initial major flow of customers. The result of this layout is that movement patterns are confused and dissipated. This dissipation takes two major forms. First, users tend to penetrate the circulation spaces which run at right angles to the initial dominant direction of flow to a limited degree only, in order to minimize the extent to which they have to retrace their steps. As a result, stalls located towards the middle of these south-west–north-east runs attract less custom. They are frequently vacant or used as storage: in effect, they constitute 'dead spots'. Second, once users veer from the initial dominant movement direction into a lateral channel, they tend not to penetrate too far: away from the entrances there is little incentive to explore in an east–west direction, and dead spots again occur towards these edges of the market.

'Dead' edges In Ngau Tau Kok market there are areas which are 'dead', in the sense that they are not used for market purposes. Frequently, this results from the fact that many of the stalls are designed so that the display of goods can take place

on one side only: the sides and backs are blank walls. Some stalls thus face on to the blank walls of adjacent stalls. Further, the market is inwardly orientated in the sense that outer walls, which could be used as selling areas orientated towards street-related pedestrian flows, are blank. A third form of dead spot results from the escalators, which take up large amounts of space: in effect, the sides of these create blank walls in the market. This is particularly problematic on the ground floor, where one of the main entrances to the market faces directly on to the blank side of an escalator: if potential customers cannot *see* a well stocked and busy market from the road as they pass by, they are less likely to enter the market. Finally, the amount of circulation space relative to selling areas is excessive, particularly on the first floor.

Dead areas such as these have an extremely negative effect on market performance. They constitute areas of non-activity and create an overall impression of commercial failure; they therefore repel customers, and stalls in the vicinity of these areas are identifiably and differentially disadvantaged. By constituting an inefficient use of market infrastructure they also affect rentals negatively: the capital and operating costs of building and running the market are frequently reflected, in one form or another, in stall rentals; and the less the spread of these costs, the higher the unit payments.

Length of stall rows Most of the uninterrupted selling runs in Ngau Tau Kok market are relatively short. As a result, it is not possible to generate the concentrations of customer flow which maximize selling opportunities for stall-holders. Furthermore, the frequent interruptions provided by cross-channels diffuse and disperse customer flows. This factor is exacerbated by the fact that in many cases stalls are not arranged in interfacing rows, but face on to the backs or sides of other stalls.

Circulation space The circulation space between stalls in Ngau Tau Kok market varies from approximately 2·5 to 3·75 metres. This is excessively wide. The result is that customers moving between rows of interfacing stalls cannot search on both sides of the row at the same time, and tend to move along either one side of the passage or the other. As a result, there is an identifiable pattern whereby one side of the run operates more effectively than the other. Further, as stated, excessive circulation space detracts from a feeling of intensity and 'busy-ness' which is integral to the commercial success of a market. Excessive amounts of space have also been left around escalators and entrances, and this contributes further to the over-scaled and under-used feel of the market.

Significantly, sellers in this market have made various efforts to overcome the problems of design and layout. Attempts have been made to reduce the circulation spaces by extending selling surfaces outwards in front of the stalls; selling surfaces made from cardboard and baskets have been extended around the blank sides of stalls; sellers have established themselves in the large circulation spaces; and on the edges of the market, selling outwards to street-related pedestrian movement occurs.

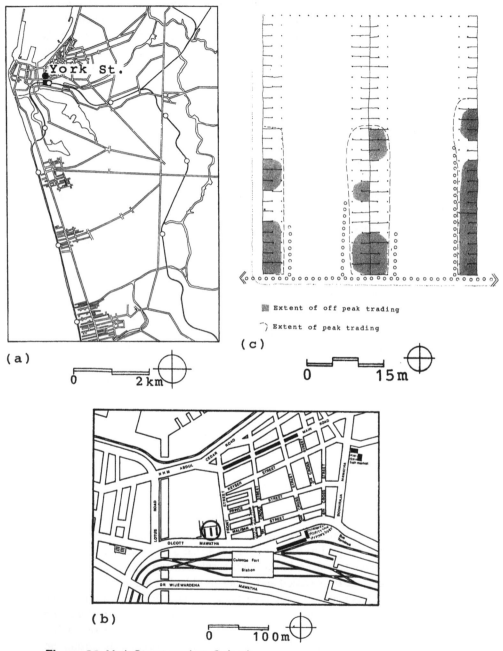

Figure 36 York Street market, Colombo
(a) Colombo
(b) Local area map
(c) York Street market

138

York Street market, Colombo

This market has been established by the municipality on the edge of the central office and commercial area of Colombo. It was originally set aside to accommodate hawkers who had been moved from street locations in the central office area. It consists of four rows of wooden kiosks on an unsurfaced lot.

This market demonstrates a number of layout problems: most particularly, it is negatively affected by changing intensities of customer flow during the day. One problem is that the layout takes the form of two culs-de-sac, each 50 metres in length. Customers entering the market are forced to retrace their steps as there is no exit on the north edge of the market and no connecting circulation channel between the two culs-de-sac. The consequence is that few customers are prepared to penetrate very far from the pavement: most stalls at the north end of the market have been abandoned.

This problem is compounded by the fact that the circulation space is excessively wide – 14 metres between interfacing stalls. This gives the market an empty and barren feel, it disperses the already scanty flows of customers, and makes it almost impossible for customers to search on both sides of the rows of stalls simultaneously. Customers are therefore discouraged still further from penetrating into the market.

Both these factors above result in the market responding poorly to changes in intensity of customer flow. At periods of peak intensity most stalls at the southern end of the market are in use. At off-peak periods, however, the already scanty customer flows are reduced to such an extent that many kiosks close down. Because stalls have been allocated to particular traders on a semi-permanent basis, however, the market is unable to contract simultaneously. Certain traders are left isolated and cut off from the remaining market activity on the southern end. There is little or no incentive for consumers to penetrate into the market to investigate these and they are, therefore, economically severely disadvantaged.

More successful market layouts

Several markets exhibit more positive layout characteristics. The markets described below illustrate different dimensions of the layout issue.

Sining South market, Taipei

Sining South market is an older fresh-produce market specializing exclusively in fruit and vegetables. It consists of two sheds which are closely aligned with surrounding roadways. The sheds intersect to create an L-shaped market, with three of the outer edges facing directly on to roads. Market infrastructure is very simple: selling surfaces are simple concrete plinths, the floor is concrete, and there are communal water facilities. There is a space of approximately 0·7 metres behind the plinths for the sellers to stand in, and between each pair of stalls there is a space of approximately 0·5 metres to allow sellers to move out from behind the stalls.

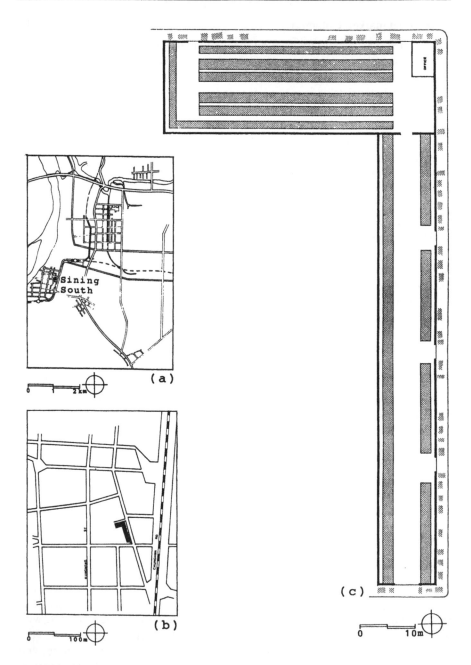

Figure 37 Sining South market, Taipei
(a) Taipei
(b) Local area map
(c) Sining South market

140

The market, which is intensively used, exhibits a number of positive features. First, the entrances to the market are appropriately placed in relation to stall layout and the natural direction of customer movement. Entrances are located at either end of both sheds and prevent cul-de-sac situations. By entering either of the sheds and moving unidirectionally towards an exit, customers are exposed to the majority of the stalls. Further, customers who would otherwise be moving along the pavement in an east–west or north–south direction, have the choice of passing through the market with only minor deviations from their original route. The longest (80 metres), east–west orientated shed has three entrances which prevent the emergence of dead areas by allowing the flow of customers through the shed to be supplemented at various points. They also create exits for people who do not wish to walk the entire length of the building.

Second, the concrete selling surfaces are orientated to the natural direction of customer flow. All stalls have equal access to passing flows of customers and as a result problems of vacant stalls, or excessive competition for more favourable locations in certain parts of the market, are avoided. The length of selling runs is sufficient to create a sense of continuity, and vibrancy. Cross-circulation passages occur at approximate intervals of 16–18 metres. This allows concentration of activity while still allowing customers to shift easily from one aisle to the next.

Third, the relatively narrow width of circulation space welds stalls on both sides into a cohesive market unit, while still allowing easy customer movement. The central passage in the larger (east–west orientated) shed is appropriately wider, since there are more stalls in this section, and it can be expected to attract more customers. Significantly, however, the 5-metre-wide passage has been narrowed slightly by the fact that the stall-holders have extended their displays into the passage. In the smaller (north–south orientated) shed, the central passage is 2 metres wide and subsidiary passages 1 metre wide. A larger space of just under 4 metres is allowed inside the entrances to accommodate the concentrations of customers which occur at these points. The space between adjacent pairs of stalls (0·5 metres) is narrow enough not to disrupt the continuity of selling runs but at the same time is sufficient to allow stall-holders to move in front of the stall if necessary.

Fourth, the layout offers considerable flexibility in terms of selling opportunities. The simplicity of the infrastructure allows easy expansion or contraction. The relatively wide pavement outside the market allows market overflow (and particularly opportunities for very small operators) to be accommodated at little or no cost: small sellers benefit from the concentration of customers generated by the built section of the market.

Pagoda Street market, Singapore

The Pagoda Street market in Singapore, which houses several hundred operators, provides an example of a situation where market layout has been spontaneously determined. The market, located in the old Chinatown area, operates day and

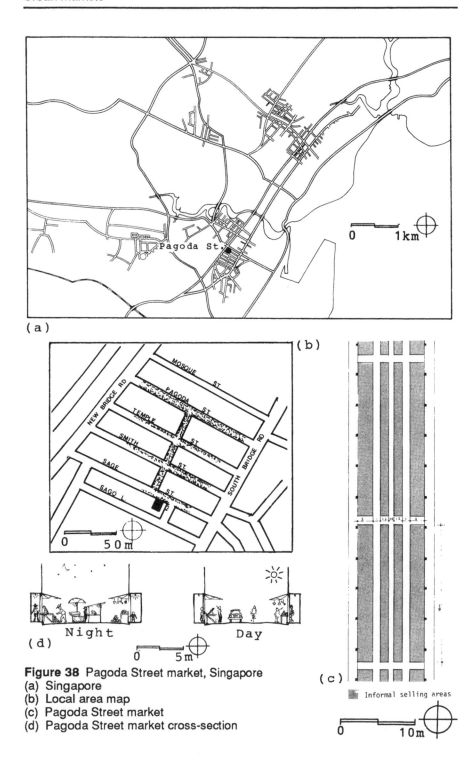

Figure 38 Pagoda Street market, Singapore
(a) Singapore
(b) Local area map
(c) Pagoda Street market
(d) Pagoda Street market cross-section

night. Stalls, made essentially of canvas hangings over bamboo struts, are erected daily by operators and at night fluorescent light tubes are tied to the struts and electrical connections run to the houses and shops flanking the street. Owners charge the stall-holders for the use of the facility.

Market configuration is essentially linear, following the alignment of the street. Access to the market can be gained from both ends of the street, as well as via a number of crossroads that intersect with it. As a result, customer flows are channelled and concentrated to the benefit of the sellers: all stalls have relatively equal access to passing customer flows, and 'dead' areas or overcongested areas do not occur. Consequently, customers are able to view the goods on display and compare prices and quality of the goods with minimum effort and minimum deviation from their dominant directional route.

In the middle section of the street, cohesive selling runs between cross-passage breaks are approximately 18–20 metres long. This ensures continuity, while allowing customers easy directional changes to other aisles. Stalls immediately adjacent to the colonnades are interrupted less frequently by cross-passages, as the general intensity of selling in this part of the street and behind the colonnades is less. Circulation space is tightly scaled and reflects the need to slow down pedestrian movement while avoiding overcongestion that may discourage potential customers from visiting the area. Passages of 1 metre have been left, both for the main through-routes and for the cross passages. A positive feature of the stall layout in Pagoda Street is that pedestrians have the choice either to move through the market along the central passages, or alternatively to avoid the market entirely and to walk along the pavements under the colonnades.

Additionally, the market is flexible in terms of its response to changes in the use of the street over the day. During daylight hours, the two middle rows of stalls are removed and vehicular movement is accommodated. The stalls on the edges of the road orientate themselves away from the road and sell to pedestrians moving along the pavement under the colonnade. In the evening, when demand for the street as a vehicular route is diminished, the central rows of stalls are erected and the street becomes entirely pedestrianized. In this way, the market is able to expand and contract without losing its cohesiveness and without developing the 'dead' areas so often associated with contraction.

Market infrastructure

No blueprints exist which determine the level at which market infrastructure should be provided in a particular context. Two factors are important, however. First, market and stall infrastructure should not be so specialized that it cannot be adapted to changes in volumes of different kinds of goods sold in a market. Second, the cost of infrastructure provision should not exclude the smaller and more intermittent sellers, as a prime purpose of a public markets policy is to accommodate these operators. The three examples below illustrate these issues.

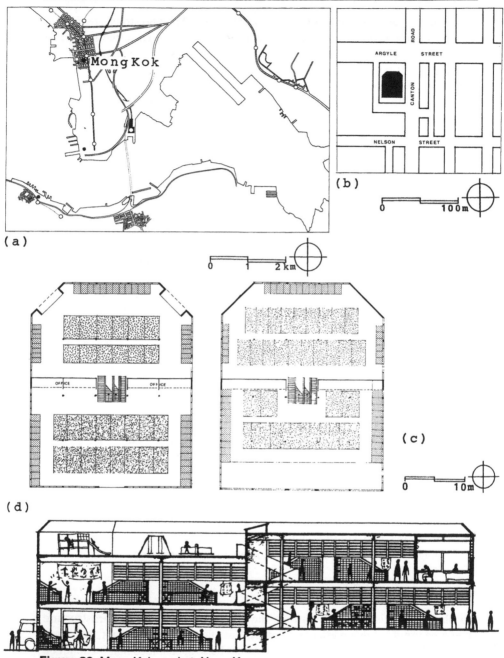

Figure 39 Mong Kok market, Hong Kong
(a) Hong Kong
(b) Local area map
(c) Mong Kok market
(d) Mong Kok market cross-section

144

Mong Kok market, Hong Kong

This is a recently completed retail market building, selling fruit, vegetables, meat, chicken, and fish. It is considered a showpiece by the local administration. A significant feature is the attempt (albeit a poor one) to combine retailing with recreational activity: the latter takes the form of children's swings and slides located on the roof of the market. The market contains 3750 square metres of selling space, arranged on five levels. There are 139 stalls, of which seventy-one are 'mini' fruit-and-vegetables stalls, and the rest, specialized stalls.

The level of infrastructure provided is high and specialized. Floors are tiled and drained and stalls and walls are also tiled. There is little natural lighting (ventilation slits occur above the level of the stalls) and this necessitates additional artificial lighting at ceiling level and over the individual stalls. There are four basic stall types in the market:

1. *Mini-stalls* are intended for relocated street hawkers. These are used for fruit and vegetables, and measure 1 by 1·5 metres. They have tiled sides and backs and some shelving. In better-located parts of the market, stall-holders have extended the stall infrastructure out into the passageway (through the use of boxes, packing cases, and baskets), thus increasing their display area.
2. *Poultry stalls* are much larger: they are 3 metres wide by 5 metres deep, and contain space for cages and for de-feathering machines. The market contains twelve poultry stalls arranged back-to-back in a double row, with a 1-metre service alley between them.
3. *Meat stalls* are 2·5 metres wide by 5 metres deep. They are provided with a cutting surface, hanging facilities, taps and basins, and storage and cooling facilities. Sixteen stalls are provided, arranged in two rows, back-to-back, with a service alley in between.
4. *Fish stalls* are 2 metres wide by 5 metres deep. There are thirty-five of these stalls: two rows on the third level of the market and two rows on the fourth level. Movement between floors in the market is provided by means both of stairs and escalators which parallel each other. There are toilet facilities and communal storage spaces. All stalls other than the mini-stalls are provided with electricity and water meters and stall-holders meet these charges individually.

The level and nature of infrastructure provided in this market give rise to four major problems. First, stalls other than the mini-stalls are so highly specialized that it is difficult to use them for the sale of goods other than those for which they were designed. The occurrence of vacant fish stalls on the upper levels is partly due to the extremely poor design and layout of the market, and partly to the fact that the stalls cannot adequately be used for any other purpose.

Second, stall rentals are high. Rent for the mini-stalls in 1983 was HK$120 (£10·88) a month: while this is low in absolute terms, it is nonetheless excessively high for many small street-hawkers. The specialist stalls were rented for between

HK$500 (£45·33) and HK$1000 (£90·67) a month and are allocated to the highest bidder. Rents of this level are high by Hong Kong standards and completely exclude the smaller and more fragile seller. The consequence of these factors is that an intense 'illegal' street market has developed immediately adjacent to the formal market.

Third, the market incurs large losses despite the high rentals. At full occupancy, some 50 per cent of stall-holders (those in the mini-stalls) generated HK$8400 (£762) a month, and the specialist stalls some HK$52500 (£4760) a month: total annual income from rental was thus HK$730,800 (£66,266). However, conservatively assuming the capital cost of the building to be HK$17 million (£1.5 million) and the opportunity or interest cost of the finance to be 10 per cent, income fails to meet even half the cost of interest payments.

Finally, the high level of infrastructure, while ensuring a hygienic and easily maintained market, nevertheless fails to create an atmosphere of vibrancy, intensity, and variety which is the key to commercial success and which is usually found in street markets. The spread of the market over four levels makes access to the upper levels difficult. The escalators themselves are an expensive item of infrastructure and create large areas of 'dead' space. The wide passages between the stalls and the large size of the stalls themselves reduce the intensity and efficiency of space use in the market, to the detriment of both the commercial attractiveness of the market and the convenience of the customer.

Sinsheng North market, Taipei

This is a recently built, retail fresh-produce market building, selling meat, chicken, fish, fruit, vegetables, clothing, and cooked food. The general level of infrastructure provided is high, and in many respects it suffers from the same problems as the Mong Kok market. However, one positive aspect of the market is the design of stall infrastructure, which is relatively simple and flexible: stall provision is therefore cheaper than in a market such as Mong Kok, and can more easily be adapted to changes in use.

The basic stall unit consists of a low, tiled concrete plinth. There are four variations of the basic stall:

1. Fruit and vegetable stalls are 1·5 metres wide by 2·5 metres deep and have a stepped front which allows for vertical displays of produce. These lend themselves to easy expansion, using more informal materials such as boxes.
2. Fish stalls are also 1·5 metres wide by 2·5 metres deep, but have slightly sloping stainless-steel surfaces which are guttered. This maximizes opportunities for display, while ensuring that potentially noxious runoff is directed into the main drainage system of the market and thus rapidly removed.
3. Poultry stalls are 1 metre wide by 2·5 metres deep. They take the form of an open-topped box with tiled stalls on three sides and cage bars on the fourth side and on top.

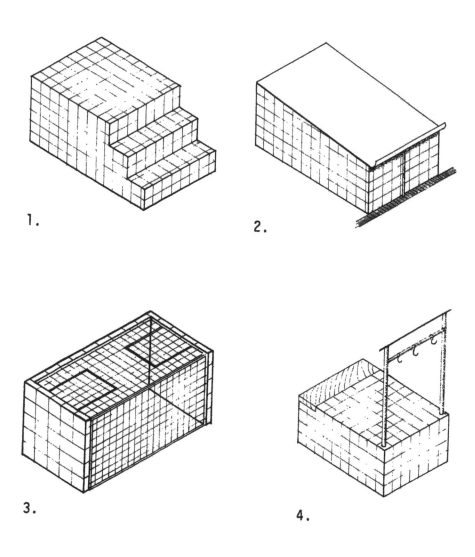

1.

2.

3.

4.

Figure 40 Sinsheng North market, Taipei

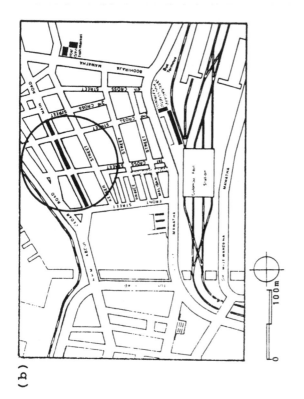

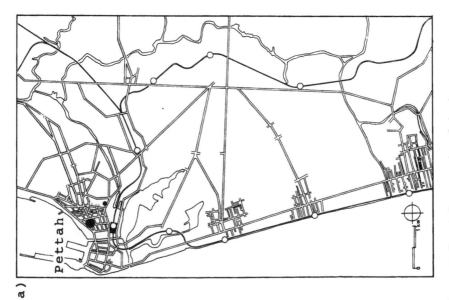

Figure 41 Pettah Street market, Colombo

(a) Colombo
(b) Local area map

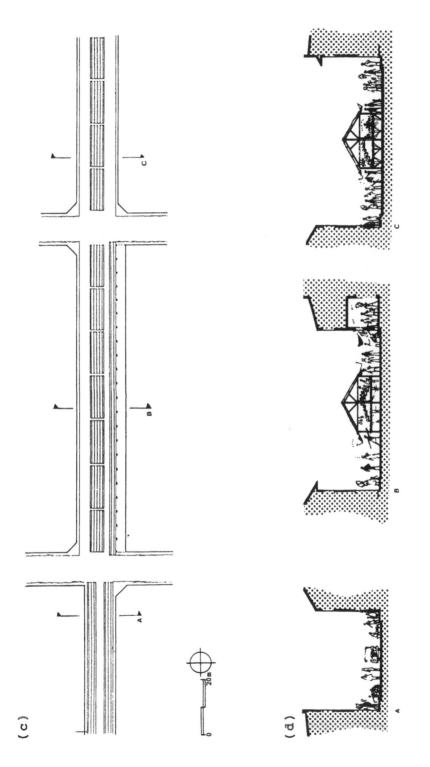

(c) Pettah Street market
(d) Pettah Street market cross-section

4. Meat stalls are 1·5 metres wide and 2 metres deep. They contain a flat surface with an inserted wooden chopping board. A light steel frame holds meat hooks and lights.

Pettah Street market, Colombo

This municipally constructed system of open-sided covered stalls represents a more formal street market. It retails primarily clothes, cloth, ornaments, and household goods. It is a highly successful market, with a simple and appropriate level of infrastructure. Adequate selling and display space is provided cheaply and flexibly and there is adequate shelter from rain and sun. Steel poles support a corrugated iron roof and wood is used to provide backs, sides, and shelving for individual stalls. Stalls are approximately 1·5 metres wide and 3 metres deep and are back-to-back. In some cases, stall-holders have taken over two adjacent stalls and have removed the partitioning between them. Stall-holders generally sell from in front of their stalls, although in some cases stalls have been 'hollowed out' to create walk-in kiosks: in other cases, stall fronts have been extended through the use of canvas overhangs.

Continuous runs of stalls are some 15 metres in length (with each run containing ten stalls on each side) with small cross-passageways to enable consumers to switch aisles. In many cases, stall-holders have extended canvas across these passageways for user comfort and have opened their display surfaces so as to sell to customers walking along both the front and the sides of the stall row. In this way, continuous shelter is provided and 'dead edge' conditions on the sides of the end stalls are avoided.

Bibliography

Bromley, R. and Gerry, C. (eds) (1979) *Casual Work and Poverty in Third World Cities*, Chichester: John Wiley.

Dewar, D. and Watson, V. (1981) *Unemployment and the Informal Sector: Some Proposals*, University of Cape Town: Urban Problems Research Unit.

International Labour Office (1972) *Employment, Incomes and Equality. A Strategy for Increasing Productive Employment in Kenya*, Geneva.

Kuhn, A. (1972) *Scope for Establishing a New Fruit and Vegetable Wholesale Market in Bangkok, Thailand*, Bangkok: FAO Regional Office.

Le Brun, O. and Gerry, C. (1975) 'Petty producers and capitalism', *Review of African Political Economy* 3: 20–32.

Moser, C. O. N. (1978) 'Informal sector or petty commodity production: dualism or dependence in urban development?', *World Development* 6 (9–10): 1041–64.

—— (1984) 'The informal sector reworked: viability and vulnerability in urban development', *Regional Development Dialogue* 5 (2): 135–83.

Sanyal, B. S. (1988) 'The urban informal sector revisited', *Third World Planning Review* 10 (1): 65–83.

Sethuraman, S. V. (1981) *The Urban Informal Sector in Developing Countries*, Geneva: International Labour Office.

Tokman, V. (1978) 'An exploration into the nature of informal–formal sector relationships', *World Development* 6 (9–10): 1065–75.

Weeks, J. (1973) 'Does employment matter?', in R. Tolley, E. de Kadt, H. Singer, and F. Wilson (eds) *Third World Employment: Problems and Strategy*, Harmondsworth: Penguin.

—— (1975) 'Policies for expanding employment in the urban informal sector of developing countries', *International Labour Review*, January.

Index